MW00843999

For a listing of recent titles in the
Artech House Radar Series,
turn to the back of this book.

MIMO Rada

Theory and Applica

MIMO Radar

Theory and Application

Jamie Bergin
Joseph R. Guerci

ARTECH
HOUSE

BOSTON | LONDON
artechhouse.com

Library of Congress Cataloging-in-Publication Data
A catalog record for this book is available from the U.S. Library of Congress.

British Library Cataloguing in Publication Data
A catalog record for this book is available from the British Library.

ISBN-13: 978-1-63081-342-0

Cover design by John Gomes

© **2018 Artech House**

10 9 8 7 6 5 4 3 2 1

Contents

4

MIMO Radar Applications 81

5

Introduction to Optimum MIMO Radar 125

6

Adaptive MIMO Radar and
MIMO Channel Estimation   149

7

Advanced MIMO Analysis Techniques   169

8

Summary and Future Work   215

Preface

It has been over 10 years since the introduction of the concept of MIMO radar and waveform diversity. During this period there has been a plethora of research activities and publications extolling the many potential benefits and pitfalls of this new area. In this book we take stock of the many developments and begin to identify where practical benefits have been achieved in real-world radars, and where work remains to flesh out remaining benefits. Beginning with MIMO radar and waveform diversity fundamentals, including the tradeoff between signal-to-noise ratio (SNR) and signal-to-interference-plus-noise ratio (SINR) and hardware/software impacts, the book moves to those areas for which measurable benefits have been achieved in actual radar systems. Examples include a production X-band radar. The latter demonstrated the ability to transform a two phase center antenna into a four virtual phase center array capable of simultaneous monopulse angle estimation and clutter cancellation. The second half of the book is then dedicated to the latest cutting edge research in MIMO and waveform diversity including optimal and adaptive MIMO, advanced "STAP on transmit", and next generation multifunction RF including simultaneous radar and communications. Much of the material is based on our own firsthand experiences and we have tried to present it in a very accessible manner targeted for a diverse audience.

This book is geared to both active radar researchers and practicing radar engineers and managers interested in the benefits of incorporating MIMO and waveform diversity techniques, as well as multifunction RF. A basic mathematical understanding of radar and signal processing is assumed, as well as an appreciation for real-time software and radar hardware (though this is not essential). It is our hope that after spending some time with the book readers will have an understanding of the basic theory of MIMO radar, waveform diversity, and multifunction RF systems. They will understand the state of the art in MIMO radar and will be provided with an understanding of when and where MIMO techniques can produce benefits for real-world systems and missions.

This book is the result of many years studying MIMO radar technology and working to apply the technology in real-world radar settings. The work was performed with the help and support of many people. First we would like to acknowledge our collaboration with Dr. John Pierro of Telephonics Corporation. His insight into microwave radar hardware was critical to finding a practical implementation of the technology for GMTI radar. We would also like to acknowledge our early collaborations with Dr. Marshal Greenspan. He provided numerous key insights into MIMO radar and the history of the subject that led to many of the ideas we present in this book. The work in this book stems from research that was performed at Information Systems Laboratories, Inc. in collaboration with Paul Techau, John Don Carlos, and David Kirk. These folks were instrumental in helping us develop our understanding of the subject and in developing the tools needed to analyze our MIMO radar ideas and implementations. In particular, the site-specific modeling and simulation techniques described in this book were pioneered by Paul Techau. Our understanding of the modeling and simulation approach described in Chapter 7 is a result of our many years of collaboration with Paul. In some cases in Chapter 7 we have adopted notations we learned from Paul for describing simulated radar signals. We also acknowledge valuable collaborations with other past and present ISL employees including Brian Watson, Doss Halsey, Chris Hulbert, Steve McNeil, Linda Fomundam, Pei-hwa Lo, Katsumi Ohnishi, Chris Teixeira, Guy Chaney, and Joel Studer.

Finally, we have set up a website with additional MIMO radar material including color versions of some of the graphics in this book. A link to this website can be found at www.islinc.com. We will periodically add content to this web site including MATLAB tools that implement some of the models discussed in this book.

1

Introduction

Multiple input, multiple output (MIMO) radar is an emerging technology that has the potential to significantly improve radar remote sensing performance in a number of important application areas including airborne surface surveillance and over-the-horizon radars. It has recently received significant attention in the research community as evidenced by the hundreds of journal, conference, and workshop papers on the topic appearing in the open literature over the past 10 to 15 years [1–3]. The hype surrounding MIMO radar, however, has come with a fair amount of skepticism. Like most new technologies, MIMO radar is not a cure-all for every radar problem. In fact, MIMO radar is not well-suited to certain radar modalities and can lead to significant performance degradation if not used judiciously.

Unfortunately, the benefits of MIMO radar have sometimes been overstated, which has created a cloud of controversy that has generally slowed its adoption for radar modes and applications where it could have great benefit [4, 5]. In this book we attempt to present a fair analysis of MIMO radar technology in a way that clearly highlights both its benefits and pitfalls. Our hope is that this book will give radar engineers the analysis tools needed to include

MIMO techniques in their trade studies, and when appropriate, in their final system designs.

The concept of a MIMO system is not new. MIMO techniques have experienced great success in other radio frequency systems, most notably in wireless communications. The underlying physics that enable both radar and communication systems to benefit from MIMO techniques are generally the same; however, the performance metrics and implementation approaches are quite different. In communications systems MIMO antennas enable improved channel capacity in complex propagation and scattering environments dominated by multipath propagation. In particular, capacity can be increased over single input, single output (SISO) wireless communications when there is significant *spatial* variation in multipath propagation. MIMO in practice greatly increases the odds that one or more of the transmit-receive paths is not in a fade and can support good-quality communications. For further details on MIMO communications the reader is referred to the rich body of literature on the subject [6].

Since most radar applications assume direct line of sight (DLoS) propagation, the advantages of MIMO radar must be somewhat different than wireless applications, which are almost always non-line of sight (NLoS) (i.e., via multipath propagation). Indeed the advantages of MIMO radar are generally more subtle and will be discussed in detail throughout this book. We briefly summarize the potential benefits to radar here:

- MIMO radar techniques can be used to synthesize virtual spatial channels or adaptive degrees of freedom (DoF). This is particularly important for compact radar applications where the number of separate transmit/receiver channels is severally limited. A real-world example involving a production X-band radar is provided in Chapter 4.

- MIMO radar provides an efficient method for broadening or spoiling the transmitter beam pattern since the full two-way transmit-receive pattern resolution can be restored in the receiver via MIMO signal processing techniques [7].

- MIMO clutter channel estimation techniques provide an effective means for rapid detection and mitigation of strong clutter discretes [8].

• Optimum MIMO techniques that jointly optimize both the transmit and receive DoF can be used to maximize radar performance [7].

A majority of the MIMO radar research presented in the literature focuses on the signal processing and waveform design aspects of the technology; however, MIMO radar is fundamentally an antenna technique that extends the concept of a multichannel receive antenna or phased array to a multichannel transmit aperture. The distinguishing feature of a MIMO system is that it is intentionally designed to produce a spatially and temporally varying antenna pattern. This is typically accomplished by exciting a multiport, multiaperture antenna with a waveform or temporal response that varies among the antenna inputs. This can have significant implications in real-world systems where metrics such as antenna gain and transmitter efficiency must be considered when trying to meet a wide range of system performance requirements including maximum detection range, thermal performance, and power efficiency.

Unfortunately, simple MIMO analysis that ignores these important antenna performance metrics can lead to impractical designs with little hope of actually working in real-world systems. Alternatively, as we will show in this book, accounting for these realistic hardware effects can actually lead to MIMO implementations with improved cost, size, weight, and power (C-SWAP) properties. In some cases we will show that MIMO systems actually result in better C-SWAP hardware solutions than competing approaches based on traditional radar techniques.

Traditional radars employ a single waveform such as a radio frequency (RF) carrier modulated by a baseband pulse. The radio frequency signal is amplified and fed into a single-port antenna with a fixed antenna pattern such as a parabolic reflector or planar slotted waveguide antenna. This type of transmitter results in illumination that produces a scaled version of the input waveform at any point in the far field of the antenna. The actual scaling of the waveform will depend on the antenna radiation pattern. The advent of active electronically scanned antenna (AESA) architectures has changed this traditional single-antenna paradigm; however, in practice the end result is typically the same. That is, the AESA is used to generate a single far-field pattern with a waveform envelope that does not vary from location to location in the far field. However, the AESA hardware does offer the flexibility to generate much more interesting space-time waveforms including those needed to support MIMO radar techniques.

MIMO radar is a space-time waveform diversity technique that represents a significant departure from a traditional radar antenna. A MIMO radar produces a spatially diverse transmit waveform or illumination of the scene. In fact, if one were to measure the MIMO pattern in the far field they would find that the transmitted radar waveform actually varies when measured at different locations in space. It is important to understand that this includes the actual waveform envelope, not just the waveform scaling due to a fixed antenna pattern.

As an example, a MIMO system might use input waveforms that are common linear frequency modulated (LFM) waveforms, perhaps with different chirp slopes, but the composite waveform as measured in the far field will typically look nothing like an LFM waveform. This happens because the spatial diversity of the transmitter results in a far-field waveform that is the coherent sum of all the input waveforms. In this way, the MIMO system imparts a spatial encoding of a scene which, under the right conditions, can be decoded and exploited to improve both target detection and location performance.

The MIMO radar concept is shown in Figure 1.1 and typically involves independent waveforms transmitted from a number of independent antenna apertures as indicated by the arrows. The different shades of the arrows indicate a unique waveform. Each waveform illuminates the target and is reflected back to the receiver where the coherent sum of the waveforms is received on each of the system antennas. We note that, in general, the antennas used to transmit do not have to be the same as those used to receive. As we will show later in the book, MIMO techniques can be used to synthesize a fully populated antenna array by transmitting and receiving on sparse transmit and receive arrays. We will show that under the right conditions this can lead to more economical antenna designs than traditional fully populated antenna arrays.

The MIMO configuration shown in Figure 1.1 is typically called a coherent MIMO system where the transmit apertures are all co-located on a single tower or platform (e.g., ship or airframe). This

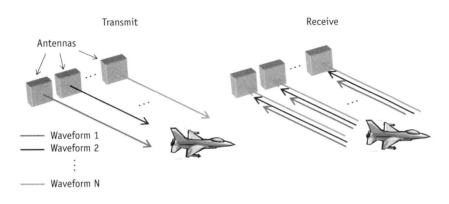

Figure 1.1   MIMO radar concept.

is in contrast to a distributed MIMO system where the transmit apertures are located on widely separated platforms and form a wide-area multistatic sensing network with transmitters and receivers often separated by tens or hundreds of kilometers. This book focuses primarily on the analysis and design of colocated MIMO systems. One of the major differences between a colocated and distributed MIMO system is that we can typically assume that the target scattering is coherent among the individual waveform returns in the colocated system since they are closely spaced (order of meters) and the aspect angle to the target does not change appreciably from antenna to antenna.

Receiver design is an important aspect of any MIMO radar system. In general, a MIMO receiver will involve multiple receive apertures, each employing a bank of receive filters at the output each matched to one of the MIMO radar transmit waveforms. An example MIMO receiver is shown in Figure 1.2. In this way the MIMO receiver can separate the radar returns produced by each transmit aperture and make them available for processing. An interesting and important implication of this architecture is that the MIMO receiver allows for the transmit pattern to be formed in the receive signal processor. Unlike a traditional single aperture system, this allows for offline scanning of the transmit pattern after all the radar signals have been collected. We will show that, under the right conditions, this property can be exploited to improve bearing estimation performance as well as improve target detection in environments dominated by clutter.

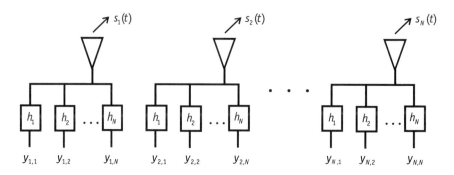

Figure 1.2   Filter h_n is matched to transmit signal $s_n(t)$ and has low correlation with all other signals.

In this book we provide analysis tools for MIMO radar systems that radar engineers can use to jointly optimize the MIMO waveform and receiver designs to maximize system performance. As discussed above, our analysis considers practical hardware considerations. We also develop and focus on radar performance metrics that not only highlight the benefits of MIMO radar, but also clearly point out the limitations and pitfalls. Further, this book focuses on the real-world aspects of the technology including limitations related to realistic hardware and interference environments.

A key aspect of our presentation is that we leverage experimental data, analysis, and performance results derived from and actual airborne X-band MIMO radar jointly developed by Information Systems Laboratories and Telephonics. A MIMO mode was incorporated into the Telephonics RDR-1700 radar system shown in Figure 1.3. This MIMO system is one of the first that addresses the various cost and design constraints of a real-world production radar system. Examples of MIMO operations over both water and land are provided that help illustrate the significant benefits of this exciting new technology. Additionally, we use this system to demonstrate that many of the practical challenges associated with MIMO radar can be overcome through appropriate hardware selection and judicious mode design.

The book is organized as follows. Chapter 2 provides the background radar signal processing prerequisites that will be used to develop the MIMO radar models and techniques described later in the book. This includes discussions of radar signals and waveforms,

Figure 1.3   Telephonics RDR 1700 radar system. This radar was modified to incorporate an advanced MIMO mode using a low-cost, low-SWAP hardware upgrade. Picture from Telphonics product website (www.telephonics.com/imaging-and-surveillance-radar).

matched filters, space-time beamforming, and Doppler processing. Chapter 3 provides an introduction to MIMO radar theory including key concepts such as the virtual array and waveform orthogonality. This chapter also provides important analysis that shows the fundamental trade-offs between MIMO and traditional radars for operations in both noise-limited and interference-limited environments. Chapter 4 shows examples of how the MIMO theory and models can be applied to specific radar applications including ground moving target indication (GMTI) radar and over-the-horizon (OTH) radar. Chapter 5 introduces advanced optimum MIMO radar theory and shows how to jointly optimize the MIMO waveforms and processing algorithms for a number of important applications including target detection in clutter and target identification. Chapter 6 shows how to extend the optimum MIMO radar theory to more realistic conditions where the MIMO radar channel is unknown and must be estimated on-the-fly (in situ) using both transmitter and receiver adaptivity. Chapter 7 presents modeling and simulation techniques for analyzing the real-world performance of MIMO systems. Finally, Chapter 8 provides a discussion of areas for future research.

References

[1]   Bliss, D. W., and K. W. Forsythe, "Multiple-Input Multiple-Output (MIMO) Radar And Imaging: Degrees of Freedom and Resolution," presented in *Conference Record of the Thirty-Seventh Asilomar Conference, Signals, Systems and Computers*, 2003.

[2]   Robey, F. C., S. Coutts, D. Weikle, J. C. McHarg, and K. Cuomo, "MIMO Radar Theory and Experimental Results," in *Conference Record of the Thirty-Eighth Asilomar Conference on, Signals, Systems and Computers*, 2004, pp. 300–304.

[3]   Bliss, D. W., K. W. Forsythe, S. K. Davis, G. S. Fawcett, D. J. Rabideau, L. L. Horowitz, et al., "GMTI MIMO Radar," presented at the *2009 International Waveform Diversity and Design Conference*.

[4]   Daum, F., and J. Huang, "MIMO Radar: Snake Oil or Good Idea?" *IEEE Aerospace and Electronic Systems Magazine*, Vol. 24, 2009, pp. 8–12.

[5]   Brookner, E., "MIMO Radar Demystified and Where It Makes Sense to Use," presented at the *2013 IEEE International Symposium on Phased Array Systems & Technology*.

[6]   Hampton, J. R., *Introduction to MIMO Communications*, Cambridge, UK: Cambridge University Press, 2013.

[7]   Guerci, J. R., *Cognitive Radar: The Knowledge-Aided Fully Adaptive Approach*. Norwood, MA: Artech House, 2010.

[8]   Bergin, J. S., J. R. Guerci, R. M. Guerci, and M. Rangaswamy, "MIMO Clutter Discrete Probing for Cognitive Radar," presented at the *IEEE International Radar Conference*, Arlington, VA, 2015.

2

Signal Processing Prerequisites

In this chapter, we review some of the key signal processing prerequisites for MIMO radar in particular and modern adaptive radar in general. While it is assumed the reader has some familiarity with these topics, they are nonetheless provided here as both a refresher and means for establishing nomenclature.

In Section 2.1, we review basic radar signals and the concept of orthogonal waveforms. We adopt a very abstract and useful vector-space definition of a transmit waveform that encompasses any and all adaptive *transmit* DoF. These adaptive DoF (ADoF) could for example include fast-time complex modulation, polarization, and spatial ADoF. Thus, a transmit signal can be quite large and complex in general.

In Section 2.2, we review the concept of matched filtering for both additive white and colored (structured) interference. Matched filtering is fundamental to both conventional radar and of course MIMO radar.

In Section 2.3, we introduce the basics of multichannel beamforming. MIMO radar makes heavy use of the spatial DoFs in achieving its goals. In later chapters, it is shown how an effective virtual increase in spatial DoF can be achieved using MIMO techniques. Since separate RF channels can dramatically increase radar costs, complexity, and size, weight, and power (SWAP), any techniques that synthesize virtual spatial DoF can be quite valuable, as discussed in Chapter 1.

Finally, in Section 2.4, we review Doppler processing as a pulse-to-pulse (as opposed to intrapulse) phase modulation that can occur naturally through relative radar-target motion or artificially through intentionally induced modulation, such as the case with Doppler division multiple access (DDMA) MIMO radar.

2.1   Radar Signals and Orthogonal Waveforms

All real-world signals are bandlimited; that is, their spectral content is bounded. For such finite norm signals it is possible to represent any continuous signal with a discrete set of samples as proven by the Shannon sampling theorem [1]. An extremely convenient and useful method for representing this set of samples is as a vector. For example, consider the continuous and bandlimited signal $s(t)$. The sampling theorem says it can also be represented, without error, by a suitably chosen set of its samples (i.e., $\{s(t_1), s(t_2), \ldots s(t_N)\}$). This in turn can be represented in vector form as

$$s(t) \rightarrow \mathbf{s} = \begin{bmatrix} s(t_1) \\ s(t_2) \\ \vdots \\ s(t_N) \end{bmatrix} \tag{2.1}$$

where by convention the vector is denoted as a boldfaced letter font. This nomenclature is very flexible. For example, if we wished to process independent signals in some joint fashion, we could simply stack the vectors to create a higher dimensional vector. Let $\mathbf{s}_1$ and $\mathbf{s}_2$ denote two N-dimensional vectors, the concatenated vector $\mathbf{s}$ would be $2N$ dimensional and have the form

$$\mathbf{s} = \begin{bmatrix} \mathbf{s}_1 \\ \mathbf{s}_2 \end{bmatrix} \tag{2.2}$$

A fundamental concept in linear algebra is the notion of a vector in some abstract vector space. Figure 2.1 illustrates the case for

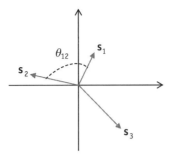

Figure 2.1　Illustration of the concept of a vector in 2-D space.

three vectors in a two-dimensional (2-D) vector space. Of course, the power of linear algebra is the ability to abstract this to an arbitrary number of dimensions, even infinity.

One can see from Figure 2.1 that vectors have both an amplitude (length) and direction. Vectors that lie along the same direction are called *collinear*, while those that are at right angles to each other are said to be *orthogonal* [2]. One of the most fundamental operations in linear algebra is the dot product, which by definition is given by

$$\mathbf{s}_1 \bullet \mathbf{s}_2 = |\mathbf{s}_1||\mathbf{s}_2|\cos\theta_{12} \tag{2.3}$$

where $\bullet$ denotes the dot product operation, $|\mathbf{s}|$ denotes the magnitude (length) of the vector $\mathbf{s}$, and θ_{12} is the angle between the two vectors as illustrated in Figure 2.1. From the properties of the cosine function it is clear that the dot product is zero when two vectors are orthogonal. This property will be critical to all of MIMO radar processing.

2.2　Matched Filtering

Consider receiving a radar signal $\mathbf{y}$ that consists of a desired target return (echo) $\mathbf{s}$ corrupted by additive random (white) noise $\mathbf{n}$; that is

$$\mathbf{y} = \mathbf{s} + \mathbf{n} \tag{2.4}$$

A fundamental radar signal processing question is "How should $\mathbf{y}$ be processed to maximize the desired signal return while simultaneously minimizing the influence of the additive noise?"

Fortunately the concept of the dot product can be used to answer this question.

Let **w** represent the filter we wish to apply to the received signal **r**. It is technically referred to as a finite impulse response (FIR) filter or more simply linear combiner [1]. What is the **w** that maximizes the signal-to-noise-ratio (SNR)? More specifically

$$\max_{\mathbf{w}} \frac{|\mathbf{w}'\mathbf{s}|}{|\mathbf{w}'\mathbf{n}|} \qquad (2.5)$$

Since the noise term is random, we need to take an additional step of maximizing the SNR *on average*. Assuming **n** is zero mean and white with a variance of σ^2 means that the expected (or average) value of the quantity $|\mathbf{w}'\mathbf{n}|$ is given by

$$
\begin{aligned}
E\left(|\mathbf{w}'\mathbf{n}|^2\right) &= E\left((\mathbf{w}'\mathbf{n})^*(\mathbf{w}'\mathbf{n})\right) \\
&= E\left(\mathbf{w}'\mathbf{n}\mathbf{n}'\mathbf{w}\right) \\
&= \mathbf{w}'E\left(\mathbf{n}\mathbf{n}'\right)\mathbf{w} \\
&= \sigma^2\mathbf{w}'I\mathbf{w} \\
&= \sigma^2\mathbf{w}'\mathbf{w}
\end{aligned}
\qquad (2.6)
$$

which implies that $E\left(|\mathbf{w}'\mathbf{n}|\right) = \sigma\sqrt{\mathbf{w}'\mathbf{w}}$. Substituting this back into (2.5) yields

$$\max_{\mathbf{w}} \frac{|\mathbf{w}'\mathbf{s}|}{\sigma\sqrt{\mathbf{w}'\mathbf{w}}} \qquad (2.7)$$

Since the magnitude of **w** (other than zero or infinity) does not affect the ratio, we can assume that it is normalized, such that $\mathbf{w}'\mathbf{w} = 1$, thus

$$\max_{\mathbf{w}} \frac{|\mathbf{w}'\mathbf{s}|}{\sigma} \qquad (2.8)$$

Since σ does not depend on $\mathbf{w}$, we see that maximizing SNR is achieved when the dot product $\mathbf{w}'\mathbf{s}$ is maximized. Since $0 \leq \cos\theta_{12} \leq 1$, the dot product is maximum when $\theta_{12} = 0$ (i.e., $\mathbf{w}$ is collinear with $\mathbf{s}$, and $\mathbf{w} = \alpha\mathbf{s}$, where α is an arbitrary nonzero scalar). One can thus clearly see where the term *matched filter* comes from.

If the noise is colored, then $E(\mathbf{nn}')$ is no longer diagonal and thus has some general form $E(\mathbf{nn}') = R$, where R is the noise covariance matrix [3]. While not derived here, a similar procedure to the above can be applied to result in the optimum colored noise matched filter, given by the famous Wiener-Hopf equation (in vector form); that is

$$\mathbf{w} = R^{-1}\mathbf{s} \qquad (2.9)$$

Thus for the colored noise case, a pre-whitening operation must first be performed, followed by a white noise matched filter matched to the whitened target signal [3].

2.3   Multichannel Beamforming

Consider the effect of a unit-amplitude, narrowband electromagnetic (EM) plane wave impinging on an N-element uniform linear array (ULA) with interelement spacing d, as depicted in Figure 2.2. In this context, the term narrowband refers to a signal whose modulation bandwidth, B, is such that $c/B \gg Nd$ [1]. This condition insures that propagation delay across the array is manifested as a simple phase shift, otherwise true time delays units (TDU) must be employed.

If we define the plane wave angle-of-arrival (AoA), θ_{o}, relative to boresight as shown in Figure 2.2, the complex envelope phasor at baseband observed at the n-th antenna element as a function of θ_{o} is

$$s_n = e^{j2\pi(n-1)\frac{d}{\lambda}\sin\theta_o}, \quad n = 1,\dots,N \qquad (2.10)$$

where λ is the operating wavelength (units consistent with d), and θ_o is the AoA in radians [4]. Note that the phase progression of a plane wave is linear across a ULA.

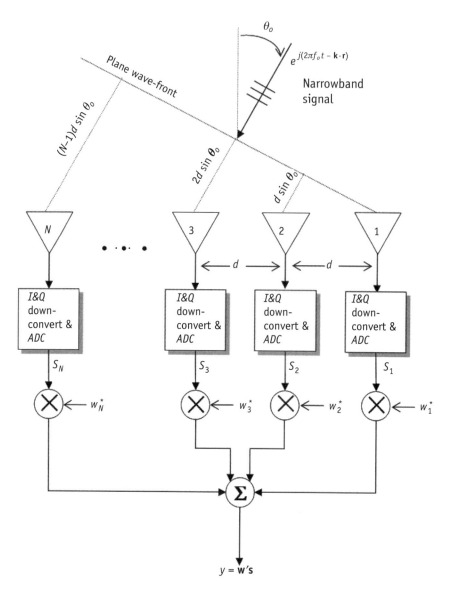

Figure 2.2   Uniform linear array beamformer. (After [4].)

By introducing multiplicative complex weighting factors, w_n, in each receive channel of the array (as shown in Figure 2.2), the output response can be maximized for any desired AoA. More specifically, let y denote the scalar beamformer output defined as

$$y = \sum_{n=1}^{N} w_n^* s_n = \mathbf{w}'\mathbf{s} \tag{2.11}$$

where * denotes complex conjugation, the prime ′ denotes vector complex conjugate transposition (i.e., Hermitian transpose [2]), and the vectors $\mathbf{s} \in \mathbb{C}^N$ and $\mathbf{w} \in \mathbb{C}^N$ ($\mathbb{C}^N$ denotes the space of N-dimensional complex vectors) are defined as

$$\mathbf{s}(\theta_0) \overset{\Delta}{=} \begin{bmatrix} s_1 \\ s_2 \\ s_3 \\ \vdots \\ s_N \end{bmatrix} = \begin{bmatrix} e^{j0} \\ e^{j2\pi\frac{d}{\lambda}\sin\theta_0} \\ e^{j2\pi(2)\frac{d}{\lambda}\sin\theta_0} \\ \vdots \\ e^{j2\pi(N-1)\frac{d}{\lambda}\sin\theta_0} \end{bmatrix} \tag{2.12}$$

and

$$\mathbf{w} = \begin{bmatrix} w_1 \\ w_2 \\ w_3 \\ \vdots \\ w_N \end{bmatrix} \tag{2.13}$$

To maximize the response of the beamformer to a plane wave arriving at an AoA of θ_0, we have the following elementary optimization problem:

$$\max_{\{\mathbf{w}\}} |y|^2 = \max_{\{\mathbf{w}\}} |\mathbf{w}'\mathbf{s}|^2 \tag{2.14}$$
$$\text{subject to} |\mathbf{w}|^2 = \text{constant} < \infty$$

where the constant gain constraint is imposed to insure a finite solution. Since $\mathbf{w}'\mathbf{s}$ is simply an inner (or dot) product of the two nonzero norm vectors $\mathbf{w}$ and $\mathbf{s}$ [3], Schwarz's inequality can be applied (i.e., $|\mathbf{w}'\mathbf{s}|^2 \leq |\mathbf{w}|^2|\mathbf{s}|^2$), with equality if (and only if) the vectors are colinear. This yields the result

$$\mathbf{w} = \kappa\mathbf{s} \tag{2.15}$$

where κ is a scalar chosen to satisfy the normalization constraint. Note that this is simply the white noise matched filter result we previously derived.

This result is intuitive inasmuch as it states that the optimum beamformer applies phase corrections to each channel to compensate for the time delays associated with the plane wave traveling across the array. More specifically, at the nth channel the beamformer forms the product $w_n^* s_n \propto e^{-j\alpha_n} e^{j\alpha_n} = 1$, thereby canceling the phase term. The beamformer thus *coherently integrates* the signal outputs from each channel. Without this compensation, destructive interference would occur with a commensurate decrease in output signal strength.

An important and fundamental limitation of linear beamformers is that they will also, in general, respond to signals arriving from other angles. This can lead to many practical problems, as strong unwanted signals from other directions can interfere with the signal of interest. To visualize this effect, consider the response of the above beamformer to plane waves arriving from $-90°$ to $+90°$ with $\mathbf{w} = \kappa\mathbf{s}$, where $\mathbf{s}$ is chosen to be a plane wave with AoA of θ_0. If we let x_n denote the output of the nth receive channel, then the total beamformer output, steered to angle θ_0, is given by

$$y = \mathbf{w}'\mathbf{x} = \kappa\sum_{n=1}^{N} x_n e^{-j2\pi(n-1)\frac{d}{\lambda}\sin\theta_0} \tag{2.16}$$

which has the form of a discrete Fourier transform (DFT) [1]. If we set $d/\lambda = 0.5$ (half-wavelength element spacing), $N = 16$, $\theta_0 = 30°$, and

$$x_n = e^{j2\pi(n-1)\frac{d}{\lambda}\sin\theta}, \quad n = 1,\ldots,N \tag{2.17}$$

while varying θ from $-90°$ to $+90°$, the beamformer response of Figure 2.3 results (with $\kappa = 1$). Note that for a ULA, the beamformer response can be obtained via a fast Fourier transform (FFT) [1]. For

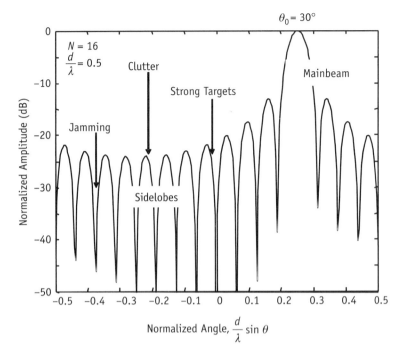

Figure 2.3 ULA beamformer response steered to $\theta_o = 30°$ with respect to boresight.

this particular example, an analytical expression for the normalized beamformer response ($|y| \leq 1$) exists and is given by (see [4])

$$|y| = \frac{1}{N} \left| \frac{\sin\left[N\pi\frac{d}{\lambda}\left(\sin(\theta) - \sin(\theta_o)\right)\right]}{\sin\left[\pi\frac{d}{\lambda}\left(\sin(\theta) - \sin(\theta_o)\right)\right]} \right| \qquad (2.18)$$

The beamformer response of Figure 2.3 has several interesting features. First, note that signals close to 30° also produce a significant response. This region is generally referred to as the *mainlobe*. The lobing structures outside of the mainlobe region are referred to as *sidelobes*. For the ULA considered, the first near-in sidelobe is approximately 13 dB down from the peak of the mainlobe [1]. The null-to-null width of the mainlobe for a ULA depends on the number of elements, N (i.e., the antenna length), and the scan angle, θ_o. It is easily obtained by setting $N\pi\frac{d}{\lambda}\left(\sin(\theta) - \sin(\theta_o)\right) = \pi$, and solving for θ_{MB}, which is the first null. The null-to-null width is thus given by

$$2\theta_{MB} = 2\sin^{-1}\left[\lambda\big/_{(Nd)} - \sin(\theta_o)\right] \tag{2.19}$$

which, for modest scan angles, can be approximated by

$$2\theta_{MB} = 2\sin^{-1}\left[\lambda\big/_{(Nd)}\right]\big/\cos(\theta_o) \tag{2.20}$$

2.4 Doppler Processing

Doppler processing in radar generally refers to filtering and/or matched filtering to produce range-Doppler (and possibly angle) output products. Moving targets will generally create a Doppler shift due to their relative motion to the observing radar. Since their velocity is not known a priori one has to set up a bank of matched filters tuned to different Doppler shift assumptions. For many typical radars this bank of matched filters can be readily implemented with an FFT, especially for nonmaneuvering targets. Even for stationary targets such as ground clutter, the motion of the radar will impart a Doppler shift. Indeed, it is this phenomenon that enables synthetic aperture radar (SAR).

Consider the effect that a Doppler shifted return propagating through a single channel, M-tap delay line filter, as shown in Figure 2.4. For a pulsed Doppler radar [5], the delay T is chosen to match the pulse repetition interval (PRI). The output of the mth tap is then given by $s_m = e^{j2\pi(m-1)\bar{f}_d}$, $m = 1,\ldots,M$, where $\bar{f}_d$ is the normalized Doppler frequency given by [5]

$$\begin{aligned}\bar{f}_d &= \frac{f_d}{\text{PRF}} = f_d T \\ &= \frac{2T}{\lambda}\left(\mathbf{v}_{tgt} - \mathbf{v}_{Rx}\right)\bullet\hat{\mathbf{i}}_{Rx}\end{aligned} \tag{2.21}$$

where PRF is the pulse repetition frequency (PRF $= 1/\text{PRI} = 1/T$), f_d is the Doppler frequency of the point target in units of $1/\text{time}$ (e.g., Hz), $\mathbf{v}_{tgt}$ and $\mathbf{v}_{Rx}$ are the target and receiver velocity vectors (Cartesian coordinates), $\hat{\mathbf{i}}_{Rx}$ is a unit direction vector pointing from the receiver to the target ($|\hat{\mathbf{i}}_{Rx}| = 1$), λ is the operating wavelength,

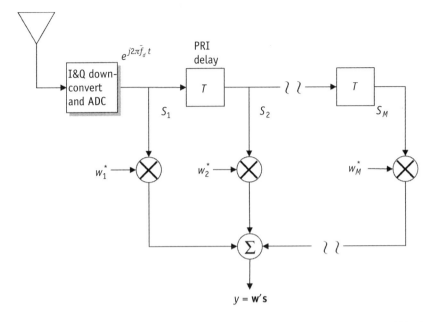

Figure 2.4 Uniform tapped delay line linear combiner for processing Doppler shifted returns. Note that the combiner output has the same mathematical form as the ULA beamformer. (After [4].)

and • denotes the vector dot product. Note for a fixed PRF, the unambiguous Doppler region for complex sampling based on the Nyquist criterion is from −PRF/2 to + PRF/2 [5]. Thus, the unambiguous normalized Doppler region is

$$-0.5 \le \overline{f}_d \le +0.5 \tag{2.22}$$

Note that (2.21) is only valid (in general) for the monostatic case (transmitter and receiver collocated).

As with the optimum ULA beamformer, an optimized Doppler filter response can also be constructed by a judicious selection of the complex weighting factors w_m, $m = 1, \ldots M$, in Figure 2.4. Let

$$\mathbf{s} = \begin{bmatrix} S_1 \\ S_2 \\ \vdots \\ S_M \end{bmatrix} = \begin{bmatrix} 1 \\ e^{j2\pi \overline{f}_d} \\ \vdots \\ e^{j2\pi(M-1)\overline{f}_d} \end{bmatrix} \tag{2.23}$$

denote the M-tap Doppler steering vector for a signal (Doppler) of interest and $w = [w_1 \ w_2 \ \ldots \ w_M]^T$ denote the vector of weights. Then, as with the ULA case, we desire to chose **w** to maximize SINR. Since the math is identical, the optimum SINR solution is also given by

$$\mathbf{w} = \kappa R^{-1}\mathbf{s} \qquad (2.24)$$

where R is the $M \times M$ total noise covariance matrix and κ is a scalar constant that does not affect SINR.

References

[1] Papoulis, A., *Signal Analysis*, New York: McGraw-Hill, 1984.

[2] Strang, G., *Introduction to Linear Algebra*: Wellesley, MA: Wellesley Cambridge Press, 2003.

[3] Van Trees, H. L., *Detection, Estimation and Modulation Theory. Part I.* New York: Wiley, 1968.

[4] Guerci, J. R., *Space-Time Adaptive Processing for Radar,* Second Edition, Norwood, MA: Artech House, 2014.

[5] Richards, M. A., *Fundamentals of Radar Signal Processing,* New York: McGraw-Hill, 2005.

3

Introduction to MIMO Radar

This chapter provides an introduction to the key principles of MIMO radar. We begin with the development of a MIMO radar signal model. We provide a summary of some of the key properties that differentiate MIMO systems from conventional radars. We conclude with the development of key performance metrics and some basic results that highlight some of the key benefits of MIMO. We will show that the MIMO antenna configuration provides additional virtual spatial antenna channels that can be used to improve radar performance when judiciously implemented. In particular, the virtual channels can provide additional DoF to improve adaptive interference mitigation. We will also show that the virtual antenna elements effectively extend the antenna aperture leading to opportunities for improved bearing estimation accuracy.

The MIMO processing approach considered in this chapter was introduced in Figure 1.2. A matched filter for every transmitted waveform is applied to the signals received on each channel. In the figure, the total number of transmitted waveforms is equal to the number of receive channels. In general, the number of transmit waveforms and receive channels can be different. We will begin, however, for the case when they are the same. The output of the matched filters provide the MIMO data vector that will be processed to cancel interference and detect targets. Thus we wish to develop a signal model that represents the vector output of all the matched filters.

We will begin with a common MIMO signal model that assumes the waveforms are truly orthogonal. This model leads to a relatively straightforward signal model that fits nicely with traditional covariance models that have been used for receive-only systems. This is the MIMO model that is most often found in the signal processing literature. In reality it is very difficult to generate and efficiently transmit perfectly orthogonal waveforms. As we will see, this is particularly true for coded waveforms such as those employed in wireless communications. Unfortunately, the commonly cited benefits of MIMO can degrade as the waveforms diverge from the truly orthogonal case. Thus, we extend the basic model to account for the nonzero cross correlation between waveforms and show how this can degrade performance. We also explain in detail from a radar systems point of view the impact of transmitting noncoherent waveforms. We also provide an assessment of the impact on computational complexity and hardware complexity when using the MIMO antenna architecture.

3.1   MIMO Radar Signal Model

We begin by considering a model for the signals received at the output of the receive antennas in Figure 1.2. A model commonly employed involves a channel matrix that accounts for the different paths traveled by the waveforms that arrive at each antenna as follows [e.g., 1, 2]:

$$\mathbf{y}(t) = H\mathbf{s}(t) \tag{3.1}$$

where $\mathbf{y}(t) = [y_1(t)\ y_2(t)\ \dots\ y_N(t)]^T$ is a vector of outputs received signals on each of the N receive antennas, $\mathbf{s}(t) = [s_1(t)\ s_2(t)\ \dots\ s_N(t)]^T$ is a vector of transmit waveforms and H is a matrix with elements of the form $v_{n_r,m_t} = \gamma_{n_r,m_t} \exp(j2\pi d\sin(\theta)(n_r + m_t)/\lambda); n_r, m_t = 0, 1, \dots, N - 1$ where v_{n_r,m_t} is the response from the m_t transmit element to the n_r receive element, γ_{n_r,m_t} is the corresponding channel response, d is the interelement spacing, λ is the operating wavelength, N is the number of antenna elements in the array, and θ is the signal angle of arrival relative to broadside of the antenna. The elements of the matrix H represent all combinations of the propagation paths between a single transmit antenna, reflected target in the far filed, and a single receive antenna.

The channel model representation of (3.1) is very common in the wireless communications literature but is not commonly used in radar analysis. This representation has a number of key benefits as we will show in later chapters. For the analysis below where we select a set of waveforms or waveform assumptions and then analyze MIMO performance this model does not generally provide any real advantage over the signal covariance models typically used in radar adaptive signal processing algorithm analysis. If we wish to move beyond the orthogonal MIMO waveforms and try to optimize the waveforms, then the channel model representation can be advantageous because, as we show in Chapter 5, the waveforms are related to the inputs via a linear relationship. In the covariance model, the waveforms are embedded in the covariance model and relate to the inputs in a nonlinear fashion, making analysis of the optimum waveform set much more difficult.

The remainder of this chapter focuses on the development of the covariance model since we are not yet interested in optimizing the waveforms but instead will be analyzing MIMO performance based on a fixed set of waveforms or waveform assumptions. We will transition over to the channel model in Chapter 5 where we develop techniques to optimize the MIMO waveforms to meet specific radar mission requirements such as interference rejection and target identification. In this chapter we will initially assume that the waveforms have negligible cross correlation so that the output of a particular filter, $h_i(t)$, only contains the filtered waveform $s_i(t)$. This assumption will hold for cases when the power of the signals due to the nonzero cross correlation is much weaker than signals of interest (targets and clutter).

For the case when the waveforms are truly orthogonal, the MIMO spatial response at the output of the matched filters is typically modeled as (e.g., [3])

$$\mathbf{v}_s(\theta) = \mathbf{v}_t(\theta) \otimes \mathbf{v}_r(\theta) \qquad (3.2)$$

where $\otimes$ is the Kronecker product and $\mathbf{v}_r(\theta)$, $\mathbf{v}_t(\theta)$ are the transmit and receive steering vectors, respectively, for a signal with angle-of-arrival θ. For a ULA the steering vectors have elements, $v_n = \exp(j2\pi nd\sin(\theta)/\lambda)$, $n = 0, 1, \ldots, N-1$, where d is the interelement spacing, λ is the operating wavelength, N is the number of antenna elements in the array, and θ is the target angle of arrival relative to broadside. It is easy to show that the elements of v_s are the same to within a scalar as the associated elements of the channel matrix H, and similarly, each represent a single path from transmit antenna to target in the far field of the antenna and back to a single receive antenna.

For a pulse-Doppler radar the MIMO space-time steering vector is given as [4, 5]

$$\mathbf{v}_{st}(\theta, f_d) = \mathbf{t}(f) \otimes \mathbf{v}_s(\theta) \tag{3.3}$$

where the vector $\mathbf{t}(f)$ is a temporal response (pulse-to-pulse) with elements given as $t_m = \exp(j2\pi mfT)$, $m = 1, 2, \ldots, M-1$, f is the Doppler frequency shift, M is the number of pulses, and T is the pulse repetition interval. The cross-correlation matrix that characterizes the correlation among the MIMO channels is given as

$$R = E\{\mathbf{y}\mathbf{y}'\} \tag{3.4}$$

where $\mathbf{y}$ is a vector that contains a single snapshot of the output of the MIMO receiver shown in Figure 1.2. For a single clutter patch or target $\mathbf{y} = \alpha \mathbf{v}_{st}(\theta, f)$ where α is a complex amplitude that is a function of the complex RCS pattern, the propagation path, and the transmit and receive antenna patterns. When the waveforms are orthogonal, the MIMO cross-correlation matrix is modeled as

$$R = |\alpha|^2 \mathbf{v}_{st}(\theta, f)\mathbf{v}'_{st}(\theta, f) \tag{3.5}$$

If we assume that radar clutter for a given range bin consists of a large number of independent radar returns, then the MIMO clutter covariance model is given as

$$R_c = \sum_{p=1}^{P_c} |\alpha_p|^2 \mathbf{v}_{st}(\theta_p, f_p)\mathbf{v}'_{st}(\theta_p, f_p) \tag{3.6}$$

where P_c is the total number of clutter patches in a given range bin. Typically, in the absence of site-specific a priori information, this model is computed by assuming a large number of clutter patches with uniform amplitudes at the same range and uniformly distributed in azimuth around the radar platform [4, 5]. The total clutter plus thermal noise covariance matrix is then given as $R = R_c + \sigma_n^2 I$, where σ_n^2 is the thermal noise variance and I is the identity matrix.

We note that, from a linear algebra point of view, the covariance model described here looks exactly like the one typically used to analyze traditional receive-only adaptive radar systems [e.g., 4]. Thus, it is often tempting to plug this new MIMO model into performance metric calculations used for receive-only adaptive systems. We caution that this can result in misleading comparisons between the MIMO and receive-only systems. As we will show below, care must be taken to accurately account for the loss in coherent transmit antenna gain when using the MIMO model. In general the complex coefficients α_p must be adjusted to account for the different antenna patterns and gains between the receive-only and MIMO systems.

We also note that the MIMO covariance model developed here will have a higher dimensionality than the receive-only system by a factor of the number of MIMO transmit waveforms. This can have a huge impact on computational complexity and system hardware complexity. We provide analysis later in this chapter comparing the added computing requirements of the MIMO system. The higher dimensionality can also impact the implementation of the adaptive signal processing algorithms. The main impact is that the higher dimensionality will require more training data to estimate the interference (clutter) covariance matrix, which is known to be a major challenge when implementing adaptive signal processing algorithms in real-world environments characterized by highly heterogeneous clutter [4]. We will revisit this issue later in this chapter.

The effects of nonzero correlation between the waveforms can be readily included in the covariance model. In general, the nonzero cross correlation will cause each radar return to introduce a small amount of unwanted noise into the received signal with some degree of correlation among the MIMO channels. We begin by writing the expression for the elements of the MIMO spatial correlation matrix as

$$
\begin{aligned}
E\{y_{n,m}y_{l,k}^*\} &= |\alpha|^2 E\left\{\sum_{i=1}^{N} v_{n,i}\tilde{s}_{i,m} \sum_{j=1}^{N} v_{l,j}^*\tilde{s}_{j,k}^*\right\} \\
&= |\alpha|^2 \sum_{i=1}^{N}\sum_{j=1}^{N} v_{n,i}v_{l,j}^* E\{\tilde{s}_{i,m}\tilde{s}_{j,k}^*\}
\end{aligned}
\tag{3.7}
$$

where $v_{n,i}$ is the spatial response when receiving on antenna n and transmitting on antenna i and $\tilde{s}_{i,j}$ is the output of waveform i from matched filter j. When the waveforms are orthogonal this model reduces to the covariance model given above. For the case when there is correlation between the waveforms we will assume the following:

$$
E\{\tilde{s}_{i,m}\tilde{s}_{j,k}^*\} = \begin{cases} 1: i = m, j = k \\ \sigma^2 : i = j, m = k \\ 0: \text{ otherwise} \end{cases}
\tag{3.8}
$$

where σ^2 is the variance of the noise introduced due to the non-zero cross correlation between the MIMO waveforms. This model accounts for the fact that the nonzero correlation introduces correlated noise among the MIMO channels. For example, cross-correlation noise that is identical to within a complex scalar will be introduced in the output of each of the filters for waveform $i = 1$ on each receive channel due to leakage through filter $j = 2$. Fortunately, this noise is correlated and can be canceled; however, as we will see, it will tend to use up the available MIMO spatial degrees of freedom that can degrade the performance benefits of the MIMO antenna.

We typically assume that the radar channels and pulses (conventional or MIMO) are processed using a linear filter (i.e., space-time beamformer)

$$
z = \mathbf{w}'\mathbf{y}_s
\tag{3.9}
$$

where $\mathbf{y}_s$ is a vector that contains the conventional or MIMO spatial channel snapshots for every pulse and will contain both targets, clutter, and thermal noise. Thus, the dimension of $\mathbf{y}_s$ is NMN_t where N is the number of receive channels, M is the number of pulses, and N_t is the number of transmit waveforms/channels.

3.2 Unique MIMO Antenna Properties

Before analyzing the performance of various MIMO processing filters $\mathbf{w}$ we will highlight two key properties of the MIMO antenna configuration. The first is the concept of transmitter antenna pattern synthesis in the receiver signal processor [1] and the second is the concept of the MIMO virtual array [2].

Unlike a traditional radar where the transmitter antenna pattern is fixed and determined by the configuration of the transmit antenna, the MIMO architecture allows for the transmit pattern to be formed in the receiver. This is easily seen by assuming that a single receive channel is employed with N_t transmit channels (multiple input, single output, or MISO). In this case, the vector $\mathbf{v}_r(\theta)$ in (3.1) is a scalar and $\mathbf{v}_s(\theta) \sim \mathbf{v}_t(\theta)$. We can select any $N_t \times 1$ spatial filter $\mathbf{w}$ that combines the transmit spatial channels to form any desired transmit pattern. This concept is shown in Figure 3.1 where there are two targets at different range delays and the MIMO system allows the transmit pattern to be steered simultaneously towards each target. In a traditional system, the transmit pattern is fixed versus range whereas the transmit pattern in a MIMO system can be range-dependent if desired. This concept is not completely new. The well-known technique of sequential lobing (e.g., [6]) involves a similar operation where multiple transmit beams are formed but separated in time. These transmit channels are then combined (as if they were collected at the same time) to compute an angle estimate. Thus, a radar employing sequential lobing is a practical example of a MISO radar system.

If we combine all the MIMO channels coherently it is possible to show that the resulting pattern is mathematically equivalent to the two-way transmit/receive pattern of a traditional radar if we use the same antenna array for transmitting and receiving. The azimuth response (antenna pattern) of the coherent MIMO processor

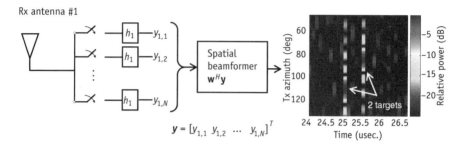

Figure 3.1 Illustration of the concept of forming the transmit pattern in the receiver in a MIMO system.

when the same antennas are used to transmit and receive and they are both steered to the same direction is given as

$$b_{MIMO}(\theta) = \left| \mathbf{v}_s'(\theta_s)\mathbf{v}_s(\theta) \right|^2 = \left| \left(\mathbf{v}_t(\theta_s) \otimes \mathbf{v}_r(\theta_s) \right)' \left(\mathbf{v}_t(\theta) \otimes \mathbf{v}_r(\theta) \right) \right|^2 \qquad (3.10)$$

where θ_s is the desired steering azimuth. Using the relationships $(\mathbf{v}_1 \otimes \mathbf{v}_2)' = (\mathbf{v}_1' \otimes \mathbf{v}_1')$ and $(A \otimes B)(C \otimes D) = AC \otimes BD$,

$$\begin{aligned} b_{MIMO}(\theta) &= \left| \left(\mathbf{v}_t(\theta_s) \otimes \mathbf{v}_r(\theta_s) \right)' \left(\mathbf{v}_t(\theta) \otimes \mathbf{v}_r(\theta) \right) \right|^2 \\ &= \left| \left(\mathbf{v}_t'(\theta_s)\mathbf{v}_t(\theta) \right) \otimes \left(\mathbf{v}_r'(\theta_s)\mathbf{v}_r(\theta) \right) \right|^2 \\ &= \left| \left(\mathbf{v}_t'(\theta_s)\mathbf{v}_t(\theta) \right) \left(\mathbf{v}_r'(\theta_s)\mathbf{v}_r(\theta) \right) \right|^2 \\ &= \left| \mathbf{v}_t'(\theta_s)\mathbf{v}_t(\theta) \right|^2 \left| \mathbf{v}_r'(\theta_s)\mathbf{v}_r(\theta) \right|^2 \end{aligned} \qquad (3.11)$$

which is clearly equal to the two-way pattern of a traditional antenna array (i.e., product of the receive and transmit patterns). So we see that coherently combining the MIMO antenna outputs can be used to scan a spatial response that is identical to a two-way antenna pattern that is well-known to have a mainbeam that is narrower than the receive antenna. Figure 3.2 shows a comparison of the spatial response for a conventional radar with eight half-wavelength spaced elements at an operating frequency of 1 GHz. Also shown is the MIMO response when orthogonal waveforms are transmitted on each of the eight antenna elements. As expected the MIMO pattern is identical to the two-way pattern of the traditional system. We will show later how this property can be exploited to

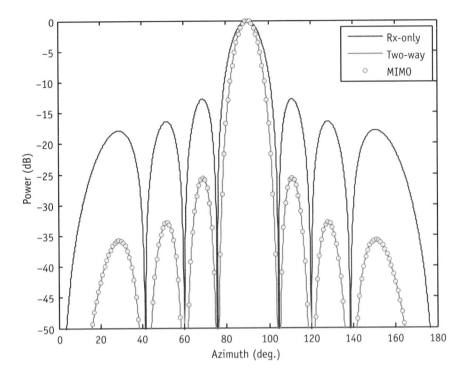

Figure 3.2 Comparison of receive-only, two-way, and MIMO antenna patterns.

improve radar functions including bearing estimation and slow target detection in airborne GMTI radar.

As we will show below, there can be serious practical implications when forming the transmit beam in the receiver. The main complication is that there is a true loss in coherent transmit gain that needs to be accounted for in the design of the MIMO radar mode and associated signal processing algorithms.

The second key property of the MIMO system is that it provides added virtual array elements with positions determined by the convolution of the positions of the receive array and transmit array [2]. As alluded to above, this will result in improved received spatial resolution. It will also result in added spatial degrees of freedom that can be used to support critical radar functions such as adaptive clutter filtering and bearing estimation (target geolocation). This concept is shown in Figure 3.3. The new array is called the MIMO virtual array and results in an antenna that is generally larger than the original receive-only system. This is illustrated by comparing the elements of the array response $\mathbf{v}_s(\theta)$ for the

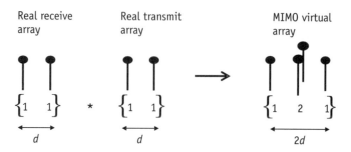

Figure 3.3 Illustration of the MIMO radar virtual array concept [2].

conventional and MIMO cases for the simple two-element antenna array shown in Figure 3.2. For the conventional receive-only case the array response is given as

$$\mathbf{v}_s(\theta) = \mathbf{v}_t(\theta) \otimes \mathbf{v}_r(\theta) = \mathbf{v}_r(\theta) = \begin{bmatrix} 1 & \exp(j2\pi d \sin(\theta)/\lambda) \end{bmatrix}^T \quad (3.12)$$

For the MIMO case it is given as,

$$\mathbf{v}_s(\theta) = \begin{bmatrix} 1 & \exp(j2\pi d \sin(\theta)/\lambda) & \exp(j2\pi d \sin(\theta)/\lambda) \\ & \exp(j4\pi d \sin(\theta)/\lambda) \end{bmatrix}^T \quad (3.13)$$

We see that the MIMO array response has four elements. The first three elements are the same as the receive-only case with the second element repeated. The fourth element of the MIMO array is the same as if a third virtual element (i.e., $n = 2$) existed in the receive-only array. Thus the MIMO configuration provides the same measurements as a larger receive-only antenna and also provides repeated or redundant measurements for some of the array elements. It's interesting to note that the virtual array has a pattern that is equivalent to an array that is twice as long as the original array but with a triangular taper.

We will show in later chapters that the larger virtual array can be exploited to improve the resolution of a MIMO system relative to a traditional receive-only system; however, care must be taken to account for the aforementioned loss in transmitter gain resulting from the MIMO configuration. We will show in Chapter 4 that the increased number of spatial channels or virtual channels can help improve system performance.

One interesting result of this property of the MIMO antenna is that it can seemingly be used to optimize the spatial response of sparse receive apertures. Since the virtual aperture antenna positions are the convolution of the receive and transmit apertures, one can imagine using a short transmit aperture with half-wavelength spacing and a larger receive aperture with spacing greater than a half wavelength in a MIMO configuration (e.g., [2]). In this case, if the transmit and receive arrays are chosen judiciously, the final virtual aperture will produce a filled array (with many redundant elements, as discussed above). This is illustrated in the patterns shown in Figure 3.4 for a system with three half-wavelength–spaced transmit antenna elements and eight receive antenna elements with uniform spacing, but with 2λ spacing between elements. We see that indeed the MIMO pattern helps reduce the grating lobes of the sparse receive aperture; however, the final pattern is identical to the two-way pattern that would result from using these two apertures for transmit and receive in a traditional non-MIMO system configuration.

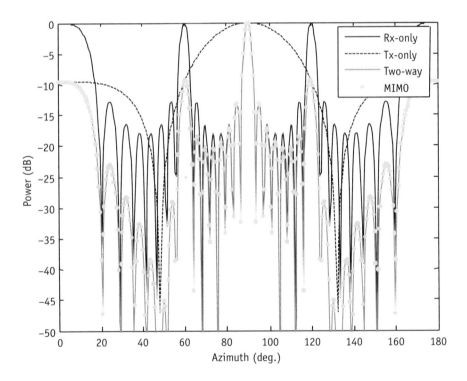

Figure 3.4 Antenna patterns for a system with sparse receive antenna.

Another interesting and potentially very useful result of the MIMO transmitter beamsteering and virtual array is that it provides a potentially more efficient alternative to traditional transmitter beam-spoiling or broadening techniques [7]. In systems with excess sensitivity the transmitter pattern can be spoiled to cover a larger instantaneous search sector allowing for multiple simultaneous received beams to be processed. This provides a straightforward trade-off between sensitivity and search area coverage rate. This is often accomplished using an appropriate phase taper across the antenna elements on transmit [7]. The downside is that the two-way antenna patterns will typically have higher sides lobes than the nonspoiled case. The MIMO antenna naturally spoils the transmit pattern in that the illumination pattern is equal to the pattern of a single channel, but as was shown above, allows for the true two-way pattern to be formed in the signal processor.

3.3   MIMO Radar System Modeling

Since a major goal of this book is to compare the differences between MIMO and conventional processing, it is important to carefully account for differences in transmit power and gain in the conventional and MIMO channels. This section follows the models presented in [8]. Another similar treatment can be found in [9]. Figure 3.5 summarizes the system models used to represent the single-pulse transmit and receive apertures for a conventional phased array radar and for a MIMO array. The major differences between MIMO and conventional antennas are (1) the MIMO transmit antenna gain will be lower due to the smaller aperture, and (2) for phased array systems employing T/R modules that distribute the total transmit power evenly across the entire aperture, the transmit power for a single MIMO channel will be reduced.

The SNR for a single receive channel on a conventional array on a single pulse is given as

$$SNR_{conv} = \frac{PG_t G_r \sigma_t \lambda^2}{(4\pi)^3 R^4 L_s kTF_n f_p} \qquad (3.14)$$

Parameter	Comment
P	Total average power
G_t	Full aperture gain
G_r	Single subarray gain
λ	Wavelength
R	Nominal slant range
L_s	System losses
k	Boltzman const.
T	Noise temp.
f_p	PRF
σ_o	Clutter coefficient
N_t	Number of waveforms
A_c	Cutter cell area (full aperture)
F_n	Noise factor

	Conventional	MIMO
Single Spatial Channel	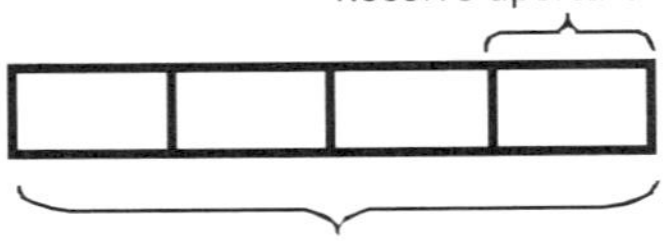 Receive aperture / Transmit aperture	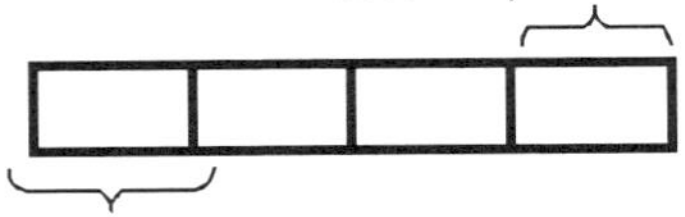 Receive aperture / Transmit aperture
SNR (Single Channel/Pulse)	$\dfrac{P G_t G_r \sigma_t \lambda^2}{(4\pi)^3 R^4 L_s kTF_n f_p} = SNR_{conv}$	$\dfrac{(P/N_t)(G_t/N_t) G_r \sigma_o \lambda^2}{(4\pi)^3 R^4 L_s kTF_n f_p} = \dfrac{SNR_{conv}}{N_t^2}$
CNR (Single Channel/Pulse)	$\dfrac{P G_t G_r \sigma_o A_c \lambda^2}{(4\pi)^3 R^4 L_s kTF_n f_p} = CNR_{conv}$	$\dfrac{(P/N_t)(G_t/N_t) G_r \sigma_o (N_t A_c) \lambda^2}{(4\pi)^3 R^4 L_s kTF_n f_p} = \dfrac{CNR_{conv}}{N_t}$

Figure 3.5 Comparison of the radar sensitivity between a conventional and MIMO radar system.

where P is the total system average power, G_t is the full aperture gain, G_r is the single-subarray aperture gain, σ_t is the target RCS, λ is the operating wavelength, R is the range to the target, L_s is system losses, k is the Boltzman constant, T is the noise temperature, F_n is the noise factor, and f_p is the pulse repetition frequency. Accounting for the different transmitter, the SNR for a single MIMO channel (single transmit channel and single receive channel) on a single pulse is given as

$$SNR_{MIMO} = \frac{\left(P/N_t\right)\left(G_t/N_t\right)G_r\sigma_t\lambda^2}{(4\pi)^3 R^4 L_s kTF_n f_p} = \frac{PG_tG_r\sigma_t\lambda^2}{(4\pi)^3 R^4 L_s kTF_n f_p N_t^2} \quad (3.15)$$

where N_t is the number of transmit subarrays employed in the MIMO array. Clearly the SNR on a single MIMO channel is lower than a single conventional array receive channel by a factor of $1/N_t^2$. The MIMO signal processing techniques discussed later in this chapter will compensate for this loss in SNR. Part of the loss will be recovered by coherently processing the MIMO channels and part will be recovered by judicious selection of the radar coherent processing interval.

Next, we consider the clutter-to-noise ratio (CNR), which, for a single receive channel on a conventional array on a single pulse, is given as

$$CNR_{conv} = \frac{PG_tG_r\sigma_o A_c\lambda^2}{(4\pi)^3 R^4 L_s kTF_n f_p} \quad (3.16)$$

where σ_o is the ground clutter scattering coefficient and A_c is the effective area of the radar resolution cell on the ground that is approximately equal to the product of the range resolution and cross-range resolution for stand-off geometries. The CNR for a single MIMO channel and a single pulse is given as

$$CNR_{MIMO} = \frac{\left(P/N_t\right)\left(G_t/N_t\right)G_r\sigma_o\left(A_c N_t\right)\lambda^2}{(4\pi)^3 R^4 L_s kTF_n f_p}$$

$$= \frac{PG_tG_r\sigma_o A_c\lambda^2}{(4\pi)^3 R^4 L_s kTF_n f_p N_t} \quad (3.17)$$

where we see that the MIMO CNR is reduced by a factor of N_t relative to the conventional array. The analysis and results presented in this book are based on these MIMO and conventional array models that account for these differences in both SNR and CNR.

3.4   MIMO Waveform Alternatives

There are many options for waveforms to support the MIMO radar concept. The desired waveform properties are

- low peak autocorrelation/range sidelobe response (PASR) [10];
- low peak cross-correlation response ratio (PCRR) [10] among the waveforms;
- Doppler tolerance for signals of interest;
- flexibility to choose an arbitrary number of waveforms and pulse widths;
- implementable using practical and economical radar hardware.

The primary objective of the MIMO waveform design is to provide a set of waveforms with very low PCRR. Typically there are three general approaches to consider: time domain waveform diversity, frequency domain waveform diversity, and code domain waveform diversity. In the following, we summarize the benefits and disadvantages of each approach and recommend the most promising waveform design approach.

Time domain waveform design. Each transmit subarray emits a signal in a unique time window. An obvious approach for radar would be to transmit a traditional LFM waveform on one pulse per subarray and cycle repeatedly through each subarray. This is a very simple approach that will likely dovetail nicely with existing system architectures; however, it will impact radar pulse repetition frequency (PRF) since the effective PRF will be reduced by a factor of the number of subarrays. This will lead to a reduction in the system area coverage by a factor of the number of waveforms used in the MIMO system. Moreover, there is an implicit assumption that the MIMO channel does not change significantly over the time period for which the TDMA probe is performed.

Frequency domain waveform design. Each subarray transmits on a different carrier frequency and separated enough in frequency to avoid mutual interference. This is also a relatively easy approach for producing orthogonal emissions from each subarray; however, it is not an efficient use of the RF spectrum and will require a total instantaneous bandwidth increase by a factor of the number of subarrays. As with the time domain case this approach would allow for the use of traditional LFM waveforms. However, the increase in instantaneous bandwidth will often make this approach impractical since higher-speed analog-to-digital converters would be required, which will in turn lead to increase system size, weight, and power requirements.

Code domain waveform design. The waveforms occupy the same time and frequency space; however, they are designed to have very low mutual correlation. This is a similar approach to modern communications systems that use code domain multiple access (CDMA) techniques. This approach represents the most efficient use of the RF spectrum; however, it is the most challenging to use in practice since achieving truly orthogonal coded waveforms in practice is difficult using finite length pulses. However, the code domain results in little impact on the system receiver hardware and does not impact the system ACR. Thus, the code domain is often selected as a starting point. However, as we will show below, even small residual cross-correlation noise can degrade the benefits of MIMO for applications such as GMTI when operating in strong clutter environments.

We highlight the challenge of developing CDMA waveforms with both low PASR and PCRR by presenting a bound on the relationship between the maximum peak autocorrelation (PASR) and the peak cross correlation (PCRR) as a function of the number of waveforms K and code length N given in [10] and is expressed as

$$PCRR \leq \frac{1}{2N-1} - \frac{2(N-1)}{(2N-1)(K-1)}PASR \tag{3.18}$$

and

$$PASR, PCRR \geq 1/N^2 \tag{3.19}$$

These bounds assume that the waveforms are normalized to have unit energy, and equation (3.19) is the result of assuming that the peak cross correlation ratio of a pair of these sequences must be greater than $1/N^2$. Other useful bounds can be found in [10, 11] and the references therein. We note that (3.19) and (3.20) represent an important bound that defines the fundamental trade-off between PCRR and PASR. Equations (3.19) and (3.20) can be used to plot a useful bound that defines the achievable PCRR as a function PASR for a given number of samples in the waveform and desired number of waveforms. Several examples are shown in Figure 3.6. These examples highlight the challenge of generating waveform sets with very low range and cross-correlation sidelobes. This is especially true when large numbers of MIMO waveforms are needed and the system time-bandwidth product is limited.

As we will show in Chapter 4, in practice it is often advantageous to employ ad hoc or hybrid waveform techniques that do not fall nicely into any of the waveform design categories above. For example, a common approach employed in airborne GMTI radar is to separate the MIMO channels in the Doppler domain using the system slow time or pulse DoF. This technique is often called DDMA MIMO (see Chapter 4) [12, 13].

We will also show that consideration of the system transmitter hardware can play a big part in the selection of the MIMO waveforms. Since the phasing and response of the transmit antenna array will be directly determined by the selection of the waveforms, care must be taken to ensure that the input waveforms provide a good impedance match to the system and do not result in excessively high voltage-to-standing-wave ratios (VSWRs) at the antenna input [14]. In general the goal will be to select waveforms that excite modes on the antenna array that are well matched to free-space propagation.

A good rule of thumb is to select waveform sets with phase responses across the antenna that result in traditional sinc-type antenna patterns. This becomes especially true for MIMO antennas where the interelement spacings are on the order of the RF wavelength, resulting in more mutual coupling between the spatial channels. For systems employing larger antenna subarrays constraints on intersubarray MIMO phase responses are less restrictive. Nevertheless, when hardware constraints are considered,

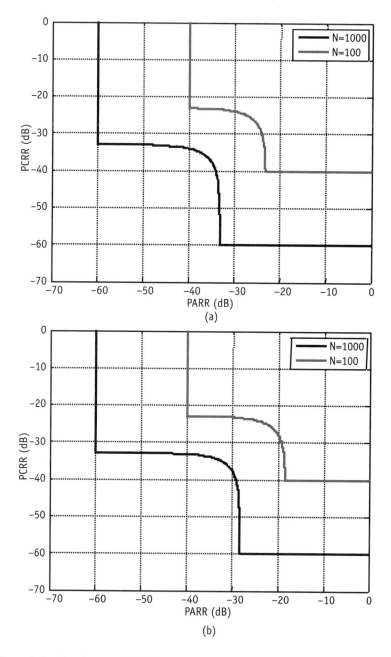

Figure 3.6 Waveform peak sidelobe and peak cross correlation bounds. (a) 2 waveforms (b) 4 waveforms. Note that PARR = PASR in these plots.

we can see how general noiselike CDMA waveforms will be very challenging to use in practice since they are likely to produce transmit antenna responses with poor matching and VSWR properties.

More detailed analysis of practical issues including VSWR will be presented later in Section 3.7.4.

3.5   MIMO Signal Processing

For detection of the target signal we employ the matched filter for the case of known colored interference plus noise, which is given as [15, 16]:

$$\mathbf{w} = \frac{R_t^{-1}\mathbf{v}_{st}(\theta, f)}{\sqrt{\mathbf{v}_{st}^H(\theta, f)R_t^{-1}\mathbf{v}_{st}(\theta, f)}} \tag{3.20}$$

where $R_t = E\{\mathbf{y}_s\mathbf{y}_s'\}$. As with traditional receive-only processing techniques, the spatial and temporal degrees-of-freedom employed will generally be chosen based on the interference environment. In practice, the ideal covariance R_t is replaced with an estimate computed adaptively using the collected radar data. Adaptive MIMO processing techniques are discussed in Chapter 5.

If we assume that the target response is $\alpha\mathbf{v}_{st}(\theta, f)$ where α is a zero-mean complex Gaussian random amplitude, then the signal-to-interference-plus-noise ratio (SINR) at the output of the space-time beamformer is given as

$$SINR = \frac{E\{|\alpha|^2\}|\mathbf{w}'\mathbf{v}_{st}(\theta, f)|^2}{\mathbf{w}'R_t\mathbf{w}} \tag{3.21}$$

If we use the filter given in (3.20) the SINR can be simplified to $SINR = E\{|\alpha|^2\}\mathbf{v}'_{st}(\theta, f)R_t^{-1}\mathbf{v}_{st}(\theta, f)$ and if we assume, without loss of generality, that the thermal noise level is unity (i.e., $\sigma^2 = 1$), then

$$SINR = SNR_{MIMO}\mathbf{v}'_{st}(\theta, f)R_t^{-1}\mathbf{v}_{st}(\theta, f) \tag{3.22}$$

If we consider the thermal noise-only case (i.e., $R_t = I$), the SINR can be expressed as

$$SINR = SNR_{MIMO}NMN_t = SNR_{conv}NM / N_t \tag{3.23}$$

where we see that for the MIMO case the SINR after processing is reduced by a factor of the number of transmit waveforms relative to the conventional array case ($N_t = 1$). To overcome this reduction in the SINR, the number of pulses must be increased by a factor of N_t in the MIMO case. Fortunately for search applications this will not impact the system area coverage rate (ACR), however, since the MIMO transmit pattern is broader by a factor of N_t due to the shorter transmit aperture. Thus, the conventional and MIMO arrays will have the same clutter-free sensitivity and ACR, allowing for a fair comparison of the performance of each to detect targets in clutter.

For many narrowband systems increasing the CPI does not seriously impact radar performance; however, care must be taken to ensure that the long CPI does not result in target signal losses due to effects such as range walk and target decorrelation arising from variations in the target RCS during the CPI. Range walk can be accommodated using more advanced Doppler processing techniques similar to SAR processing algorithms such as keystone processing [17]. In dense target environments, the longer CPI can often be beneficial because it provides improved resolution of targets that are closely spaced in Doppler. This can be especially beneficial for systems employing adaptive clutter mitigation because it will minimize the number of targets that corrupt the training data used to estimate the clutter filter weights (e.g., [18]).

For the case when the interference is not white, the comparison between the MIMO and conventional is not as straightforward and typically depends on the nature of the target response in relationship to the interference response. To demonstrate, we consider a baseline airborne GMTI system operating at L-band (1 GHz) with a 1m horizontal antenna aperture aligned with the body of the aircraft and divided into four equal sized subapertures. The system is traveling at a speed of 100 m/s and the antenna is steered toward broadside of the antenna (i.e., side-looking). The baseline conventional system uses a coherent processing interval made up of 16 pulses with PRF of 2 kHz. The target does not fluctuate during the CPI.

A number of SINR curves are shown in Figure 3.7 that highlight the differences between noise-limited and clutter-limited performance. As expected the MIMO case requires an increase in the

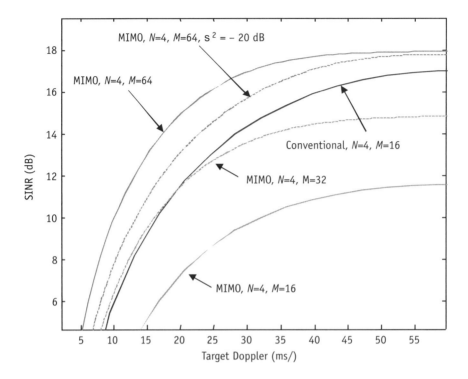

Figure 3.7 Comparison of sensitivity between a conventional antenna and various MIMO configurations. Unless noted, $\sigma^2 = 0$.

number of pulses by a factor of the number of transmit waveforms to achieved the same performance as the conventional system in the noised-limited (high target velocity) region. In this case, the MIMO system provides improved SINR in the clutter-limited region (low target velocity). This is a result of the narrower MIMO spatial response discussed above, which helps isolate slow targets from competing clutter signals. This property will be explored in more detail in Chapter 4. An intermediate case is also presented where the MIMO system has 32 pulses. In this case the MIMO system will have an ACR that is 2× greater than the conventional system and also provides better SINR in the low Doppler region. Thus we see that for clutter-limited cases the MIMO system can often provide very significant performance gains. This will be explored in more detail in Chapter 4.

Another key performance metric used to compare MIMO systems to traditional antennas is the bearing estimation accuracy.

Performance bounds for the *thermal noise* limited case are provided by the following previously published expressions [9]

$$\sigma_\theta^{MIMO} = \frac{BW}{\sqrt{2SINR}} \tag{3.24}$$

and

$$\sigma_\theta^{conv} = \frac{BW}{\sqrt{SINR}} \tag{3.25}$$

where BW is the receive antenna beamwidth. We see that given the same SNR, a MIMO system will achieve better angle estimation accuracy by a factor of $\sqrt{2}$ [9]. We will show in Chapter 4 that in clutter-limited cases the relative improvement in MIMO over conventional antennas can be even greater. Also, it is important to note that these bounds are for bearing estimation accuracy when a single target is present. In radar we are often interested in understanding accuracy when two or more targets are present or how well the system can resolve multiple targets. Given the narrower response of the MIMO system, we anticipate that the MIMO system will have improved resolution relative to conventional systems.

Typically the power output of the space-time beamformer given in (3.9) is computed and compared with a detection threshold to test for the presence of a target:

$$\left| \mathbf{w}^H \mathbf{y}_s \right| > \gamma \tag{3.26}$$

If we assume that both targets and interference in the data vector $\mathbf{y}_s$ have zero-mean complex Gaussian amplitude statistics, then the filter output will also be zero-mean Gaussian and the detection statistic will have an exponential distribution that can be used to readily compute the probability of detection and probability of false alarm as

$$p_d = e^{-\gamma/(\sigma_t^2 + \sigma_n^2)} \tag{3.27}$$

and

$$p_{fa} = e^{-\gamma/\sigma_n^2} \tag{3.28}$$

respectively, where σ_t^2 is the target average power and σ_n^2 is the interference (i.e., clutter plus noise) power. It is straightforward to show that $p_d = p_{fa}^{1/(SINR+1)}$. This relationship along with the SINR given in (3.21) can be used to compute the receiver operating characteristic (ROC) for both conventional and MIMO systems.

This relationship provides the p_d and p_{fa} for a single coherent processing interval (CPI); however, radar systems typically employ post detection logic or integration that combines the detections from several CPIs to form a final dwell-level detection. We will assume that N of M processing logic is employed. In this case a target must be detected in at least N out of M CPIs in a dwell to be declared a true detection. If we assume that the detections are independent from CPI to CPI then the following expressions can be used to compute the dwell level probability of detection, $p_d^{N,M}$, and probability of false alarm, $p_{fa}^{N,M}$,

$$p_d^{N,M} = \sum_{n=N}^{M} \frac{M!}{n!(M-n)!} p_d^n (1-p_d)^{M-n} \tag{3.29}$$

and

$$p_{fa}^{N,M} = \sum_{n=N}^{M} \frac{M!}{n!(M-n)!} p_{fa}^n (1-p_{fa})^{M-n} \tag{3.30}$$

3.6 Site-Specific Simulation Example

The performance metrics described in the previous section can be used to assess and compare MIMO performance with traditional receive-only processing techniques. In this section we present a mini case study for an airborne radar engagement to show how the MIMO capability can lead to improved radar performance. This simulation was first presented in [8]. Some of the results are included here with additional discussion related to the metrics discussed earlier in this chapter. The example involves a notional airborne GMTI radar system that is attempting to detect and track a slow-moving ground vehicle over time. We will provide more details about GMTI radar modes in Chapter 4, but this example is meant to show how all the MIMO properties described earlier in this chapter have the potential to improve radar mission performance.

While this example is not an exhaustive analysis of radar tracking performance from detection through tracking, it does provide a sufficient level of realism that demonstrates how the metrics described above can be used to compare traditional and MIMO system performance. This example also highlights a number of key issues that arise when comparing MIMO and traditional surveillance radar systems. In particular, care must be taken when comparing the false alarm rates between the traditional and MIMO systems to account for the longer MIMO integration time.

In the simulation, an aircraft at an altitude of 3 km passes a position on the ground. The aircraft is heading due north, with the radar aim point located off to the west at a slant range of 32.5 km at the start of the simulation. The simulated airborne radar system consisted of a ULA with parameters tabulated in Figure 3.8. This represents a notional airborne radar system with parameters that

Radar Parameters

	Conv	MIMO
Spatial channels	4	4
System loss (dB)	9	9
Noise temperature (K)	270	270
Noise figure (dB)	5	5
PRF (kHz)	1.984	1.984
Bandwidth (MHz)	10	10
Frequency (GHz)	1.24	1.24
Array length (m)	1.44	1.44
Array width (m)	0.12	0.12
Average power (W)	1500	1500
Number of pulses	8	32

Figure 3.8 Radar parameters for the traditional radar 'conv' and the MIMO system. (© 2009 IEEE. Reprinted, with permission, from *Proceedings of the 2009 Aerospace and Electronic Systems Conference.*)

are similar to the L-band system described in [19]. The parameters have been modified slightly to improve the GMTI capabilities of the system. In particular, the bandwidth was increased to reduce the clutter-to-noise ratio and help resolve targets. We note that the number of pulses for the MIMO system is greater by a factor of four relative to the conventional system to account for the losses discussed earlier in this chapter. We also note that in this simulation we are assuming that the MIMO waveforms are perfectly orthogonal and that they are generated using a CDMA-like approach in fast-time (range dimension). Thus, this example represents a sort of bound on MIMO performance.

A ground vehicle was included in the simulation using available GPS position data. The GPS track is shown in Figure 3.9. The vehicle makes an initial left turn and accelerates to a consistent speed on a straight road. Note that the vehicle is moving at the beginning and end of the track. From the plot of target speed over time, target speed reduction at the turn is identifiable around track time $t = 34$ seconds. In the vicinity of the turn, a benefit provided by MIMO radar in detecting slow targets will be observed.

Target SINR was recomputed every second during the duration of the track to produce the SINR versus time shown in Figure 3.10. As expected, the MIMO system results in improved SINR for this target, which is within the mainbeam clutter region. This is the MIMO gain in SINR discussed earlier in this chapter for cases when

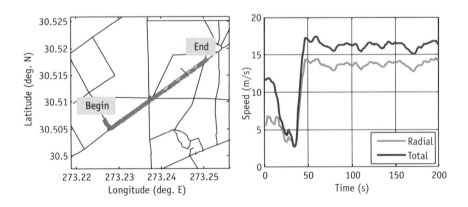

Figure 3.9   Left: Site-specific target track. Right: Target total and radial speed. (© 2009 IEEE. Reprinted, with permission, from *Proceedings of the 2009 Aerospace and Electronic Systems Conference.*)

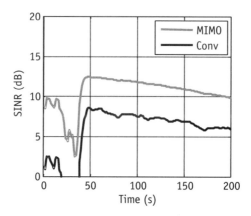

Figure 3.10   Left: Target SINR versus time. (© 2009 IEEE. Reprinted, with permission, from *Proceedings of the 2009 Aerospace and Electronic Systems Conference.*)

the radar detection problem is interference-limited (as opposed to thermal noise-limited). The target SINR values were used to generate detection performance statistics. Figure 3.11 shows probability of detection over time, assuming a probability of false alarm of $p_{fa} = 10^{-5}$ for the conventional system and $p_{fa} = (1/N_t) \times 10^{-5} = (1/4) \times 10^{-5}$ for the MIMO system. Different false alarm rates were used for conventional and MIMO performance comparison to maintain a similar false alarm density (i.e., false alarms per unit surveillance area). While MIMO and conventional radar have the same area coverage rate, more Doppler bins are present in the MIMO radar output. This leads to a larger false alarm density compared to conventional

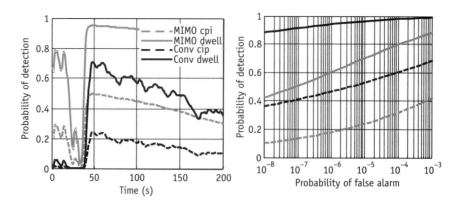

Figure 3.11   Left: Target probability of detection during simulation. Right: pd vs. pfa at time t = 59 seconds. (© 2009 IEEE. Reprinted, with permission, from *Proceedings of the 2009 Aerospace and Electronic Systems Conference.*)

radar if the false alarm rate is maintained; hence the need for a smaller MIMO false alarm rate to obtain similar false alarm densities with each system. Single-hit detection (dashed lines) and 2 of 5 dwell detection (solid lines) processes were implemented.

A comparison of target p_d and system p_{fa} at $t = 59$ seconds is also shown in Figure 3.11. We see that MIMO radar provides significantly improved detection performance over the conventional radar. From these results it is obvious that the MIMO system will provide a better capability to detect and track slow-moving targets than a traditional single-waveform system.

The performance of MIMO and conventional systems was analyzed as a function of the number of channels. Figure 3.12 shows the SINR versus time for two-, three-, and four-channel systems. In the case of MIMO the number of transmit waveforms is equal to the number of receive channels. As can be seen, the MIMO system performance is more robust to the reduction in the number of channels since both transmit and receive channels are available (i.e., for the two-channel case the MIMO system has two transmit and two receive spatial DoF). We also note that for the two-channel case the traditional array angle estimation performance will be very poor since the available spatial degrees of freedom will be required to cancel clutter. In the case of the MIMO system the extra transmit spatial DoF combined with the receive DoF will help improve the

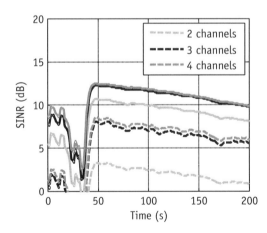

Figure 3.12 Target SINR versus time for different numbers of channels as indicated by the color in the legend. Solid lines represent MIMO radar and dashed lines represent conventional radar. (© 2009 IEEE. Reprinted, with permission, from *Proceedings of the 2009 Aerospace and Electronic Systems Conference.*)

angle estimation performance when the number of receive DoF is limited.

The azimuth angle estimation error versus time is shown in Figure 3.13 for both the MIMO and the conventional array. We see that the MIMO system provides a reduction in azimuth estimation errors of approximately a factor of two relative to the conventional array. This is due to both the effective increase in the antenna aperture (i.e., the factor given in [3.24]) and the improved SINR for slow moving targets.

This simulation example involving a realistic radar engagement geometry demonstrates how the virtual extension of the antenna aperture induced by the MIMO architecture results in better performance for detecting slow-moving targets and improved angle estimation accuracy compared to conventional radar systems. It is clear from the performance metric results presented that the tracking performance of the MIMO systems will be significantly better than the traditional system. For example, the probability of detection of the MIMO systems was close to unity whereas the traditional system probability of detection during the engagement rarely exceeded 0.6. The difference between the two systems was emphasized early in the engagement where the target radial velocity was very low. In this case the MIMO system probability of detection

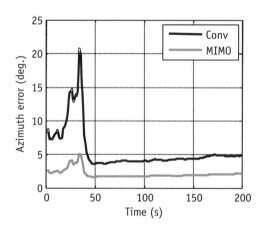

Figure 3.13   Azimuth angle RMS estimation error bound during simulation.(© 2009 IEEE. Reprinted, with permission, from *Proceedings of the 2009 Aerospace and Electronic Systems Conference.*)

was often above 0.8 whereas the traditional system probability of detection never exceeded 0.3.

3.7   MIMO Implementation Issues and Challenges

The benefits of MIMO radar come at a cost in terms increased computational requirements as well as hardware complexity. As well, the added MIMO spacial channels can lead to practical challenges when implementing adaptive clutter filter training strategies. Finally, finding a good set of MIMO waveforms that do not impact transmitter efficiency can be challenging in practice. In this section we address a number of these challenges and potential mitigation strategies. We will show in Chapter 4 that it is indeed possible to overcome many of these challenges; however, it requires a MIMO implementation that represents a significant departure from the general MIMO approach described earlier in this chapter involving quasi-orthogonal CDMA-type waveforms generated in fast-time on each channel.

3.7.1   Computational Complexity

We begin by analyzing the computational requirements of the receiver processing to support MIMO radar. We focus on the main processing blocks typically encountered in radar: (1) pulse compression, (2) Doppler processing, and (3) clutter mitigation. We typically think of the pulse compression and Doppler process as a preprocessing step prior to clutter mitigation. Thus, below we will consider these two steps separately. Basic computational complexity models are developed for each of these blocks for both MIMO and traditional radars. We will see that the preprocessing step does not increase computational complexity significantly for search-type radars when considering the number of operations required per unit search area. The clutter mitigation does, however, significantly increase the required number of operations.

We begin by developing a computational model for pulse compression for a traditional single waveform radar system. We assume that pulse compression is performed for each radar channel, N, and pulse, M. We model the number of operations required for pulse compression as

$$OPS_{pc} = NML \log_2 L \qquad (3.31)$$

where L is the number of range bins. We are assuming that the pulse compression is computed using fast convolution with FFTs and therefore will have computations proportional to $L\log_2 L$ [20]. We are mainly interested in comparing this model with a similar model for MIMO radar presented below. Therefore, we are ignoring factors due to complex multiplication and other miscellaneous factors that would be accounted for if we were concerned with the exact number of operations as opposed to the number of operations relative to MIMO.

The number of operations required to Doppler process the radar data for every channel and range bin is modeled as

$$OPS_{dop} = NLM \log_2 M \qquad (3.32)$$

where we have assumed that the Doppler processing is computed using an FFT and is proportional to $M\log_2 M$ [20]. The number of operations to preprocess the data is then the sum of the pulse compression and Doppler processing operations,

$$OPS_{pre} = OPS_{pc} + OPS_{dop} = NML \log_2 (LM) \qquad (3.33)$$

The number of operations to pulse compress the data for the MIMO radar is greater than the traditional radar since N_t additional matched filters are required behind each of the receive antenna elements, as was shown in Figure 1.2. Also, there will be N_t more pulses, as discussed earlier in this chapter, to make up for the loss in transmitter antenna gain. The model for the number of MIMO pulse compression operations is given as

$$OPS_{pc,MIMO} = N_t^2 NML \log_2 L \qquad (3.34)$$

where the factor of N_t^2 accounts for the additional matched filters and the increased number of pulses in the coherent processing interval. Again, we are assuming that the pulse compression is computed using fast convolution with FFTs and therefore will have computations proportional to $L\log_2 L$ [20].

The number of operations required to Doppler process the MIMO radar data for every MIMO channel (transmit and receive) and range bin is modeled as

$$OPS_{dop,MIMO} = N_t^2 NLM \log_2 MN_t \tag{3.35}$$

The factor N_t^2 accounts for the increased number of channels and pulses. Whereas the traditional radar required M pulses to be Doppler processed for N channels and L range bins, the MIMO systems requires MN_t pulses to be Doppler processed for NN_t channels and L range bins. Again, we have assumed that the Doppler processing is computed using an FFT and is proportional to $MN_t \log_2 MN_t$ [20]. The number of operations to preprocess the MIMO data is then the sum of the pulse compression and Doppler processing operations,

$$OPS_{pre,MIMO} = OPS_{pc,MIMO} + OPS_{dop,MIMO} = NN_t^2 ML \log_2(LMN_t) \tag{3.36}$$

We see that the number of raw operations counts for the MIMO system is significantly greater than the traditional single waveform system. For search applications this result is misleading since the MIMO system employs a longer CPI and therefore has longer to perform the operations and also covers a larger surveillance area than the traditional system because of the broader (spoiled) transmitter antenna. We can perform a more meaningful and fair comparison by computing the number of operations per second per unit surveillance area. We note that the number of operations per second is commonly referred to as floating point operations per second or FLOPS.

If we assume without loss of generality that the traditional radar employs a 1-second CPI (i.e., the pulses take one second to transmit) and the antenna beam and range swath cover 1 square kilometer, then the expression given in (3.33) is also the FLOPS/km^2. For the corresponding MIMO system with the same area coverage rate the CPI will be equal to N_t seconds and the surveillance area will be N_t km^2. Thus the number of FLOPS per square kilometer is computed by dividing (3.36) by the factor N_t^2 giving,

$$FLOPS\_norm_{MIMO} = NML \log_2(LMN_t) \ \text{FLOPS}/\text{km}^2 \tag{3.37}$$

The increase in FLOPS for the MIMO system relative to the traditional system to surveil the same area in the same amount of time is the ratio of (3.37) and (3.33) and is given as

$$fac_{pre} = 1 + \frac{\log_2 N_t}{\log_2 LM} \tag{3.38}$$

This factor does not represent a major increase in the computing requirements of the MIMO system to preprocess the data for typical values of N_t, M, and L. For example, if $N_t = 4$, $M = 128$, and $L = 1,000$ the increase in required operations is approximately 12%, which is a rather modest increase. We note that this represents a sort of worst-case example for MIMO since we have assumed CDMA-like waveforms that require additional matched filters on each channel. Other MIMO implementations like the DDMA approach discussed later in Chapter 4 will result in less impact on computing for the prepressing operations (pulse compression and Doppler processing). We will see below, however, that the increase in computing for the clutter mitigation can be quite significant and must be seriously considered when comparing the cost/benefit analysis between MIMO and traditional systems.

The computational models represented by (3.38) assume that the longer MIMO CPI does not result in significant range-walk of targets and clutter due to either ownship motion of the radar platform or target motion. This will typically be the case for MIMO implementations involving a small number of waveforms and for narrowband systems. For applications when this is not the case, more sophisticated nonseparable pulse compression and Doppler processing may be needed to avoid integration losses and clutter decorrelation effects due to range walk. In this case preprocessing akin to synthetic aperture radar processing may be needed, which could add significant computational complexity to the MIMO system to perform preprocessing of the data.

The computing model for the clutter mitigation processing is for the clutter filter weights given in (3.20). The computational complexity of this type of adaptive space-time beamformer (e.g., space-time adaptive processing [STAP]) has been studied extensively (e.g., [21]). The operations are mainly driven by the estimation of the clutter covariance matrix and subsequent inverse as well as the application of the resulting weights to the radar data. Here we will employ a rather simplified model that captures the key computations required since the goal is to compare traditional and MIMO systems as opposed to calculating exact numbers of operations. The number of operations for a baseline single-waveform system is represented as

$$OPS_{STAP} = M\left(M_f N\right)^3 + NM_f LM \tag{3.39}$$

where M_f is the number of temporal degrees-of-freedom used in the clutter mitigation processing. This number will depend on the particular STAP algorithm employed. For example, a common algorithm is multibin post Doppler STAP [4] where a small number of Doppler bins ($M_f <<M$) are used in the STAP processing. The first term in (3.39) is for the estimation of the clutter covariance matrix and computing its inverse. There are many ways to perform these operations; however, they all typically require numbers of operations proportional to the number of space-time degrees of freedom (NM_f) cubed [21]. The factor of M in the first term in (3.39) arises because we must compute a unique clutter filter for each pulse or Doppler bin. We also note that this analysis assumes a single or global set of weights is computed and applied to all range bins. The second term in (3.39) is for the application of the weights to the radar data or simply the inner product of the weights with the radar data. Each weight application or inner product requires NM_f operations and this must be repeated for each of the L range bins and M Doppler bins or pulses.

A similar expression for the MIMO STAP beamformer that accounts for the extra spatial channels due to the N_t MIMO waveforms is given as

$$OPS_{STAP,MIMO} = N_t^4 M\left(M_f N\right)^3 + N_t^3 NM_f LM \tag{3.40}$$

Again the first term is for the calculation of the weights and the second term is for the application of the weights to the radar data. We note that this expression is the same as (3.39) with $N = NN_t$ and $M = MN_t$. That is, both the number of spatial channels and number of Doppler bins (or pulses) are increased by a factor of the number of MIMO waveforms employed as discussed earlier in this chapter. Also, the second term has been multiplied by an additional factor of N_t because in the MIMO system the covariance matrix will be used to compute space-time weights for N_t unique steering directions to cover the wider surveillance swath covered by the spoiled transmitter beam. We note that we have ignored the operations to recompute the weights since this can be done using the same covariance inverse with a small and generally negligible number of operations.

As was the case with the preprocessing functions analyzed above, for search radars the MIMO system employs a longer CPI and therefore has longer to perform the operations and also covers a larger surveillance area than the traditional system because of the broader (spoiled) transmitter antenna. We can perform a more meaningful and fair comparison of the STAP computational complexity by computing the number of operations per second per unit surveillance area.

Here we will also assume without loss of generality that the traditional radar employs a 1-second CPI (i.e., the M pulse take 1 second to transmit) and the antenna beam and range swath cover 1 square kilometer, then the expression given in (3.39) is also the FLOPS/km² for clutter mitigation processing. For the corresponding MIMO system with the same area coverage rate the CPI will be equal to N_t seconds and the surveillance area will be N_t km². Thus the number of FLOPS per square kilometer is computed by dividing (3.40) by the factor N_t^2 giving,

$$
\begin{aligned}
&FLOPS\_norm_{STAP,MIMO} \\
&= N_t^2 M\left(M_f N\right)^3 + N_t NM_f LM \quad \text{FLOPS/km}^2
\end{aligned}
\tag{3.41}
$$

Taking the ratio of (3.41) and (3.39) provides the increase in operations required for clutter mitigation by the MIMO system relative to the traditional single waveform system and is given as

$$
fac_{STAP} = N_t \left(\frac{N_t M_f^2 N^2 + L}{M_f^2 N^2 + L} \right)
\tag{3.42}
$$

The second term will be close to unity for typical radar paramaters where L is often large. For example, if $N_t = 4$, $M = 128$, $M_f = 3$, and $L = 1,000$, the second term is 1.4. Thus we see that the increase in computing required for MIMO is approximately equal to the number of MIMO waveforms. This can be a very significant increase given that even the bare minimum number of waveforms needed to have a MIMO system will require a factor of two increase in the system signal processing hardware! For airborne systems with SWAP constraints, this has the potential to pose significant implementation challenges.

3.7.2 Adaptive Clutter Mitigation Challenges

It is clear from the analysis above that the increased MIMO channels lead to increased numbers of adaptive DoF that can drive up computational requirements. The increased DoF can also complicate the actual implementation and resulting performance of the adaptive clutter mitigation filters in real-world environments characterized by heterogeneous clutter [22]. The clutter mitigation weights are typically estimated using available training data in the range dimension (i.e., adjacent range bins). It is well known that the number of training data samples must be on the order of the number of DoF to ensure that estimation losses remain low. The famous result, known as Brennan's rule [23], states that when the number of training samples is twice the number of adaptive DoF, the losses will be on the order of 3 dB. This result requires stationary training data that can often be difficult to come by in real-world clutter environments characterized by severe terrain and man-made clutter returns. As an example, Figure 3.14 shows a realistic clutter map for an airborne GMTI radar generated using ISL's RFView^{TM} software [24]. This software, which is discussed in more detail in Chapter 7, uses very accurate terrain and land cover databases to produce very realistic, site-specific, clutter scenes for use in analyzing adaptive signal processing algorithms. In this case the simulated radar is flying off the coast of Southern California with its antenna mainbeam steered over land near San Diego. We see that in many regions the clutter varies drastically in the range dimension due to the varying terrain in the scene. This type of clutter heterogeneity can lead to very poor adaptive clutter mitigation performance because there is not enough training data to support the estimation of the clutter mitigation filters.

This problem has been extensively studied for traditional single waveform systems and has led to the development of a range

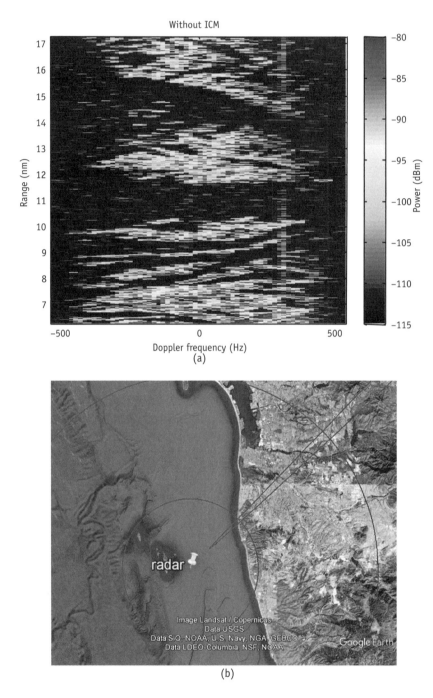

Figure 3.14 Example of a heterogenous clutter environemt generated with ISL's RFView [24] site-specific radar simulator that uses terrain and land cover databases to generate highly realistic clutter scenes. (a) Simulated clutter. (b) Scenario showing radar location, range swath, and antenna mainbeam.

of candidate solutions including reduced-DoF STAP beamformer implementations (e.g., [4]). Some of the most promising solutions involve the use of prior knowledge about the clutter environment to precondition or detrend the heterogeneous data prior to adaptive processing. These techniques all fall into the class of STAP algorithms known as knowledge-aided STAP (KA-STAP) (e.g., [25–30]). They are also sometimes called knowledge-based STAP (KB-STAP) (e.g., [31–33]).

The KA-STAP techniques can be readily applied to the a MIMO system in much the same way any baseline STAP algorithm can be applied in a MIMO system as was discussed earlier in this chapter. Once the MIMO preprocessing is completed, the extra MIMO channels can be simply viewed as added spatial DoF to include in the space-time beamformer. All that is required is a model of the spatial DoF. As was shown earlier in this chapter this is nothing more than a spatial steering vector that includes the added transmit DoF. Figure 3.15 shows a typical KA-STAP architecture [30] that would be a good candidate for use with a MIMO system. In this case the processor is termed knowledge-aided MIMO-STAP (KA-MIMO-STAP). We see that the architecture involves a knowledge processor that accumulates information about the clutter environment which is fed to the more common signal processor that incorporates this knowledge into the beamforming via the colored loading technique [27, 34]. This type of processing was found to be extremely beneficial for traditional single waveform systems operating in heterogeneous clutter environments [27]. The importance of KA processing is expected to be even more important for MIMO systems where the numbers of DoF are generally greater. We will show in Chapter 6 how MIMO techniques can be exploited to potentially overcome these challenging training issues by employing a novel clutter probing and channel estimation methodology.

3.7.3 Calibration and Equalization Issues

Calibration techniques are often required to obtain low sidelobes with an array as well as ensure good angle of arrival (AoA) measurement accuracy. However, in the case of the mitigation of ground-scattered radar energy (clutter), consideration must be given to the impact of any miscalibration on the system's ability to cancel the interference to a very low level. In the case of STAP [4],

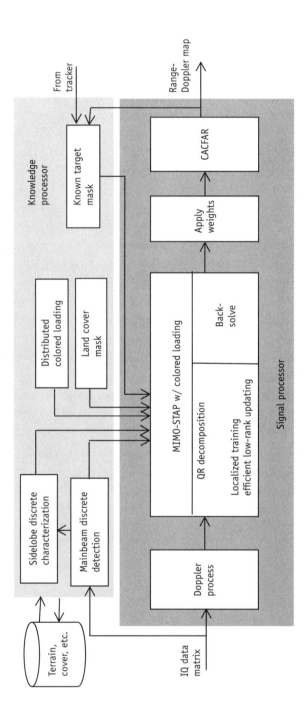

Figure 3.15 Knowledge-Aided MIMO-STAP (KA-MIMO-STAP) architecture.

the rank of the interference may be increased, thus complicating the clutter mitigation. In addition, mismatch between the assumed and true target steering vector may result in signal losses. Array calibration effects on array processing for traditional single wave-form systems have been well-studied in previous work (see, for example, Chapter 12 of [35] and references thereto).

Most systems employ an in situ calibration capability. An example is provided in Figure 3.16 where we show an antenna array with switches directly behind each antenna where a calibration signal can be injected. This signal is received and processed to remove unknown errors in each channel. We will begin by reviewing how these systems typically operate to calibrate the physical receiver channels and then discuss the challenges of calibrating the synthesized channels in a MIMO system[1].

Varying line lengths from each element to the appropriate receiver as well as other factors can result in an unknown phase

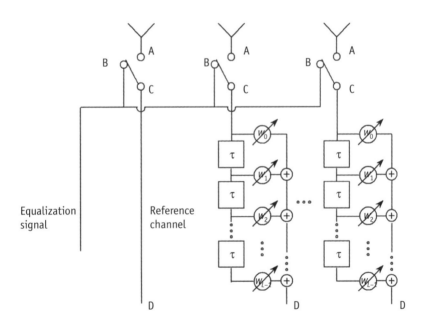

Figure 3.16 Typical array equalization approach. Each of the weights w_l is adjustable. The delay between each tap is τ. (© ISL, Inc. 2017. Used with permission.)

1. The authors acknowledge many useful discussions with Paul Techau about channel calibration and equalization techniques. Mr. Techau helped developed some of the material in this section when he was with Information Systems Laboratories, Inc.

shift and amplitude for each channel. While the line lengths may be controlled and other gain factors calibrated, temperature variations, and so forth can still result in calibration errors. Usually, the channel phase is a more significant factor. If the amplitude and phase of the equalization signal injected at B in Figure 3.16 are known, then the channel equalization would compensate for the channel gain and phase variations. However, if the line length to each element for the injected signal is not perfectly known, then the phase of the equalization signal at B will not be known. (Typically, the amplitude of this injected signal is more easily controlled.) Thus control of phase and amplitude of the uplink of the equalization signal is required or calibration errors will result.

One approach to this problem is external calibration. An external source placed at a known location can be used to determine the array response for a given AoA. This response is then used together with assumed element position locations to determine a phase calibration for each element. However, any errors in element location, multipath, or mutual coupling effects will result in errors in the resulting calibration vector. Multiple external sources may be useful in addressing this problem. If the phase for each beacon (four or more) is measured, then the position and phase of each element can be determined by solving a set of simple equations in a least-squares sense.

The level of effectiveness of this technique will depend on the level of mutual coupling and multipath, among other factors, for the platform/array combination. Onboard calibration beacons may be possible to provide in situ calibration in some circumstances, depending on the array/platform configuration. Large planar arrays with subapertures or beamformed outputs will require special consideration. Another technique for providing in situ calibration is to use clutter returns (e.g., references in [35]). Given knowledge of the platform velocity and the array orientation, the angle-Doppler spectrum of the clutter is well known [4]. This relationship can be used to refine the system calibration.

Equalization is typically needed to account for variations in the passband characteristics of the channels of the array. This is generally accomplished by injecting a wideband (matching the bandwidth of the channel passband) waveform immediately after the antenna inputs (with the antenna shut off to prevent interference from the environment) and then solving for a set of tap-delay weights (w_l) on each channel that make all the channels match a reference (typically one of the channels of the array) (e.g., [36]). This is shown in Figure 3.16.

There are several issues that can arise in the effectiveness of this approach. First, enough taps must be used to compensate for the variations among the passband characteristics of the channels in the array. Also, the reference signal must have sufficient SNR at the input to each channel (e.g., if 40 dB of clutter cancellation is desired, then the reference signal must have an SNR greater than 40 dB).

A more difficult issue to address is the impact of matching between the antenna and the channel. Consider the situation in Figure 3.16. The equalization weights are calculated to maximize the cancellation when the signal is in position B. However, because of matching issues, the channel characteristics when the switch in is position A can vary from when it is in position B. In other words, the characteristic of the channels ACD vary from the channel characteristics of the channels BCD. This will tend to limit the cancellation ratio that is achievable among the channels BCD. External equalization signals can address this problem but will be ineffective when other signals are present.

Clearly calibration and equalization of a receive-only antenna array is challenging. The MIMO architecture further complicates calibration and equalization because existing hardware architectures like the one shown in Figure 3.16 do not readily provide a way to calibrate the gain, phase, and passband characteristics of the channel traveled by each waveform. If the same hardware paths are used for transmitting and receiving then it might be possible to use the receive calibration information to model the transmit responses. Another approach would be to add the capability to transmit a calibration waveform through each transmitter and sequentially receive it on each receiver. This would provide a data set that would potentially allow all the transmit/receive channels in the system to be estimated and equalized similar to the approach discussed above using an injected calibration signal. Of course, this will add hardware complexity and still has the limitation of not being able to fully remove calibration and equalization errors due to differences and mutual coupling among the antenna elements. The impact and mitigation of MIMO radar calibration and equalization errors is an area that will require further research as MIMO systems become more widely adopted and used for radar missions involving challenging clutter mitigation requirements.

3.7.4   Hardware Challenges and Constraints

A straightforward MIMO implementation where each antenna element in the radar is used to transmit arbitrary and unique waveforms requires significantly more complex hardware than the corresponding single waveform system. The added complexity is due to the need for multiple arbitrary waveform generators and independent power amplifiers behind each channel. Fortunately, there are MIMO waveforms that work well in many applications but do not require arbitrary gain and phase control. That is, they can be implemented with a small number of gain and phase states. We will show an example of this in Chapter 4.

One major issue with MIMO systems that is not typically encountered when developing algorithms for traditional single waveform systems is that the selection of the waveforms can impact the transmit hardware. For a traditional system the selection of the radar waveform is often constrained to be constant modulus (phase-only modulation) to accommodate the use of efficient transmit amplifiers; however, arbitrary phase modulation is typically not a major concern. As we will show next, for MIMO systems, the relative phase between the waveforms can impact transmitter efficiency and have a deleterious effect on the transmitter hardware when there is mutual coupling between antenna elements in the array.

We employ a very simple two-port device model (e.g., [37–39]) to highlight the hardware interactions in a MIMO system that need to be considered when designing MIMO waveforms. The antenna model is shown in Figure 3.17. The input and output relationship is modeled using an S-parameter formulation [40, 41] as follows:

$$\begin{bmatrix} b_1 \\ b_2 \end{bmatrix} = \begin{bmatrix} S_{11} & S_{12} \\ S_{21} & S_{22} \end{bmatrix} \begin{bmatrix} a_1 \\ a_2 \end{bmatrix}$$

(3.43)

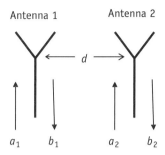

Figure 3.17 Simple two-port antenna array model.

In this model the vector $[a_1\ a_2]^T$ represents the instantaneous value of the two MIMO waveforms we wish to transmit. The parameters s_{11} and s_{22} account for how much of the input signal is reflected at the antenna input terminals. Obviously a good system will be designed such that the reflections are small with most of the applied power transmitted into free space. The parameters s_{12} and s_{21} represent the mutual coupling between the two antennas. In an ideal system these parameters will be zero, in which case the performance of each antenna will be independent. In practice there will always be some amount of coupling between the antennas and we will show next that this coupling can cause interactions between the MIMO waveforms that must be considered in any practical MIMO implementation.

We are interested in analyzing the transmitter efficiency as a function of the input waveforms. A common metric used to analyze antenna performance is the VSWR [14]. Others have also considered VSWR in the design of MIMO waveforms using measured S-parameters [42]. The VSWR metric characterizes how well an antenna or device is matched to a transmission line or other device. In practice well-matched devices and antennas lead to good power efficiency. For example, we want most of the power applied to our antenna array to transfer to the antenna and propagate into free space. Low values of VSWR indicate good matching and overall good power efficiency. The VSWR metric is defined as [14]

$$VSWR = \frac{1+|\Gamma|}{1-|\Gamma|} \tag{3.44}$$

where Γ is the reflection coefficient defined as the ratio of the output and input of a single port (e.g., $\Gamma = b_1/a_1$). It is clear that the VSWR is a function of the antenna inputs when the mutual coupling parameters s_{12} and s_{21} are nonzero. Thus, power efficiency depends on the MIMO waveforms.

The S-parameters are often modeled using an appropriate electromagnetic code (e.g., [43]) or measured in a laboratory setting [44]. Here we will use a simple but logical model given as

$$S = \begin{bmatrix} 0.2 & 0.2e^{j2\pi d/\lambda} \\ 0.2e^{j2\pi d/\lambda} & 0.2 \end{bmatrix} \tag{3.45}$$

where d is the separation between the antennas and λ is the radar operating wavelength. We are assuming a small reflection at the inputs $s_{11} = s_{22} = 0.2$, which in the absence of mutual coupling, would result in a reflection coefficient of 0.2 and VSWR = 1.5, which is considered a good operating point for a practical system resulting in 96% of the power transferred into the antenna. We will assume the same magnitude for the mutual coupling coefficients s_{12} and s_{21}; however, we assume that the phase will depend on the separation between the antennas. The assumption is that the signal input into one of the antennas will couple into the other antenna with a propagation delay equal to $\tau_c = d/c$ where c is the speed of light in free space. Under a narrowband assumption this delay leads to a propagation phase equal to $2\pi d/\lambda$ modeled as the coefficient in the exponential terms in (3.45).

The VSWR for port one as a function of the antenna array scanning angle is shown in Figure 3.18 for the simple s-parameter model in (3.45) with d = $\lambda/2$. The first observation is that when the array is steered to broadside the VSWR is low. In this case the signal arriving at port one from coupling from port two is 180° out of phase and helps cancel the reflected signal in port one. As well, the VSWR degrades gracefully as the array is scanned away from broadside. Thus, we see that array inputs that steer natural beams (e.g., sinc patterns) typically result in efficient operation of narrowband antenna arrays with half-wavelength spacing.

The VSWR for the case when two MIMO waveforms are input into the simple antenna model is shown in Figure 3.19. In this case the waveforms are simply two unit amplitude random

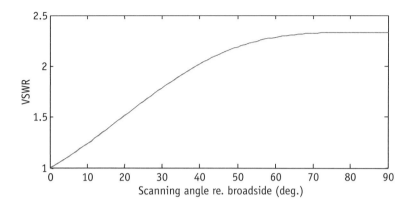

Figure 3.18 VSWR as a function of the antenna scan angle.

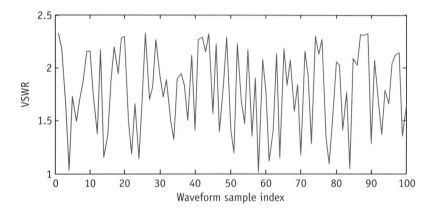

Figure 3.19 VSWR as a function of the waveform sample number for a MIMO systems employing random phase coded waveforms.

phase-coded sequences with 100 samples with phase values that are uniformly distributed on $[0, 2\pi]$. As expected, we see that the VSWR can vary wildly during the waveform and sometimes takes on a very high value. In practice these waveforms will typically result in low power efficiency as well as large reflected signals that can potentially damage the system hardware. We note that the impact of the waveforms will depend on the degree of mutual coupling between the antennas. For example, the impact will be less for higher-frequency systems (e.g., X-band GMTI radar) employing large subarrays with narrow antenna patterns than for lower-frequency systems (e.g., OTH radar) employing closely spaced antenna elements with broad patterns.

References

[1]   Bliss, D. W., and K. W. Forsythe, "Multiple-Input Multiple-Output (MIMO) Radar and Imaging: Degrees of Freedom and Resolution," *Conference Record of the Thirty-Seventh Asilomar Conference on Signals, Systems and Computers,* Nov. 9–12, 2003, Vol. 1, pp. 54–59.

[2]   Bliss, D. W. K., W. Forsythe, and G. Fawcett, "Multiple-Input Multiple-Output (MIMO) Radar and Imaging: Degrees of Freedom and Resolution," *Adaptive Sensor and Array Processing (ASAP) Workshop*, Lexington, MA, June 6–7, 2006.

[3]   Bekkerman, I., and J. Tabrikian, "Target Detection and Localization Using MIMO Radars and Sonars," *IEEE Transactions on Signal Processing*, Vol. 54, No. 10, October 2006, pp. 3873–3883.

[4]   Guerci, J. R., *Space-Time Adaptive Processing for Radar*, Second Edition, Norwood, MA: Artech House. 2014.

[5]   Guerci, J. R., *Cognitive Radar: The Knowledge-Aided Fully Adaptive Approach*, Norwood, MA: Artech House, 2010.

[6]   Lo, K. W., "Theoretical Analysis of the Sequential Lobing Technique," *IEEE Transactions on Aerospace and Electronic Systems*, Vol. 35, No. 1, January 1999.

[7]   Kerce, J., G. Brown, and M. Mitchell, "Phase-Only Transmit Beam Broadening for Improved Radar Search Performance," *Proceedings of the 2007 IEEE Radar Conference*.

[8]   Bergin, J. S., S. McNeil, L. Fomundam, and P. Zulch "MIMO Phased-Array for SMTI Radar," *Proceedings of the 2008 IEEE Aerospace Conference*, Big Sky, MT, March 2–7, 2008.

[9]   Rabideau, D., and P. Parker, "Ubiquitous MIMO Multifunction Digital Array Radar," *Conference Record of the Thirty-Seventh Asilomar Conference on Signals, Systems, and Computers*, Vol. 1, November 9–12, 2003, pp. 1057–1064.

[10] Keel, B.M., J.M. Baden, and T. Heath, "A Comprehensive Review of Quasi-Orthogonal Waveforms," *2007 IEEE Radar Conference*, Boston, April 2007.

[11] Welch, L. R. "Lower Bounds on the Maximum Cross Correlation of Signals," *IEEE Transactions on Information Theory*, Vol. IT-20, May 1974.

[12] Mecca, V., J. Krolik, and F. Robey, "Beamspace Slow-Time MIMO Radar for Multipath Clutter Mitigation," *IEEE International Conference on Acoustics, Speech and Signal Processing*, ICASSP 2008.

[13] Mecca, V., D. Ramakrishnan, and J. Krolik "MIMO Radar Space-Time Adaptive Processing for Multipath Clutter Mitigation," *Fourth IEEE Workshop on Sensor Array and Multichannel Processing, 2006*.

[14] Balanis, C., *Antenna Theory: Analysis and Design*, New York: John Wiley & Sons, 2016.

[15] Van Trees, H. L., *Optimum Array Processing*, New York: John Wiley and Sons, 2002.

[16]   Chen, C.- Y., and P. P. Vaidyanathan, "MIMO Radar Space-Time Adaptive Processing Using Prolate Spheroidal Wave Functions," *IEEE Transactions on Signal Processing*, Vol. 56, No. 2, 2008, pp. 623–635.

[17]   Perry, R. P., R. C. DiPietro, and R. Fante, "Coherent Integration with Range Migration Using Keystone Formatting," *Proceedings of the 2007 IEEE Radar Conference.*

[18]   Bergin, J. S., P. M. Techau, W. L. Melvin, and J. R. Guerci, "GMTI STAP in Target-Rich Environments: Site-Specific Analysis," *Proceedings of the 2002 IEEE Radar Conference*, Long Beach, CA, April 22–25, 2002.

[19]   Fenner, D. K., and W. F. Hoover, "Test Results of a Space-Time Adaptive Processing System for Airborne Early Warning Radar," *Proc. of the 1996 IEEE Radar Conference*, Ann Arbor, MI, May 13–15, 1996, pp. 88–93.

[20]   Brigham, E., *Fast Fourier Transform and Its Applications,* Upper Saddle River NJ: Prentice Hall, 1988.

[21]   Borsari, J., and A. Steinhardt, "Cost-Efficient Training Strategies for Space-Time Adaptive Processing Algorithms," *1995 Conference Record of the Twenty-Ninth Asilomar Conference on Signals, Systems and Computers,* Pacific Grove, CA, August 2002.

[22]   Melvin, W. L., "Space-Time Adaptive Radar Performance in Heterogeneous Clutter," *IEEE Transactions on Aerospace and Electronic Systems*, Vol. 36, No. 2, 2000.

[23]   Reed, I. S., J. D. Mallett, and L. E. Brennan, "Rapid Convergence Rate in Adaptive Arrays," *IEEE Trans. AES*, Vol. 10, No. 6, November 1974.

[24]   https://rfview.islinc.com/RFView/.

[25]   Bergin, J. S., C. M. Teixeira, P M. Techau, and J. R. Guerci, "Reduced Degree-of-Freedom STAP with Knowledge-Aided Data Pre-Whitening," *Proceedings of the 2004 IEEE Radar Conference*, Philadelphia, April 26–29, 2004.

[26]   Bergin, J. S., G. H. Chaney, and P. M. Techau, "Performance Evaluation of Knowledge-Aided Processing Architectures," *Proceedings of the Adaptive Sensor Array Processing Workshop*, MIT Lincoln Laboratory, Lexington, MA, June 6–7, 2006.

[27]   Bergin, J. S., C. M Teixeira, P. M. Techau, and J. R. Guerci, "Improved Clutter Mitigation Performance Using Knowledge-Aided Space-Time Adaptive Processing," *IEEE Transactions on Aerospace and Electronic Systems*, Vol. 42, July, 2006, pp. 997–1009.

[28]   Guerci, J. R., and E. J. Baranoski, "Knowledge-Aided Adaptive Radar at DARPA: An Overview," *IEEE Signal Processing Magazine*, Vol. 23, No. 1, January 2006.

[29]   Capraro, C., G. Capraro, D. Weiner, M. Wicks, and W. Baldygo, "Improved STAP Performance Using Knowledge-Aided Secondary Data Selection," *Proceedings of the 2004 IEEE Radar Conference*, Philadelphia, April 2004, pp. 361–365.

[30] Bergin, J. S., D. R. Kirk, G. Chaney, S. C. McNeil, and P. A. Zulch, "Evaluation of Knowledge-Aided STAP Using Experimental Data," *Proceedings of the 2007 IEEE Aerospace Conference*, Big Sky, MT, March 1–8, 2007.

[31] Capraro, C., G. Capraro, D. Weiner, and M. Wicks, "Knowledge Based Map Space Time Adaptive Processing (KBMapSTAP)," *Proceedings of the 2001 International Conference on Imaging Science, Systems, and Technology*, Las Vegas, NV, June 2001, pp. 533–538.

[32] Wicks, M., et al. "Space-Time Adaptive Processing: A Knowledge-Based Perspective for Airborne Radar," *IEEE Signal Processing Magazine*, Vol. 23, January 2006.

[33] Capraro, G., et al. "Knowledge-Based Radar Signal and Data Processing," *IEEE Signal Processing Magazine*, Vol. 23, January 2006.

[34] Hiemstra, J. D., "Colored Diagonal Loading," *Proceedings of the 2002 IEEE Radar Conference*, Long Beach, CA, April 22–25, 2002.

[35] Klemm, R., *Space-time Adaptive Processing: Principles and Applications*, London: The Institution of Electrical Engineers, 1998.

[36] Haykin, S., *Adaptive Filter Theory*, Upper Saddle River, NJ: Prentice Hall, 1996.

[37] Wang, M., W. Wu, and Z. Shen, "Bandwidth Enhancement of Antenna Arrays Utilizing Mutual Coupling between Antenna Elements," in *International Journal of Antennas and Propagation: Mutual Coupling in Antenna Arrays*, T. Hui, M. Bialkowski, and H. Lui (Guest Editors), 2010, Hindawi Publishing Corporation.

[38] Lo, K., and T. Vu, Simple S-Parameter Model for Receiving Antenna Array," *Electronics Letters*, Vol. 24, No. 20, September 1988.

[39] Haynes, M., and M. Moghaddam, "Multipole and S-Parameter Antenna and Propagation Model," *IEEE Trans. on Antennas and Propagation*, Vol. 59, No. 1, January 2011.

[40] Pozar, D., *Microwave Engineering*, New York: John Wiley and Sons, 2012.

[41] Harrington, R. F., *Time Harmonic Electromagnetic Fields*, New York: McGraw-Hill, 1961.

[42] Mecca, V., et al. "Slow-Time MIMO STAP with Improved Power Efficiency," *Conference Record of the Forty-First Asilomar Conference on Signals, Systems, and Computers*, Pacific Grove, CA, November, 2007.

[43] G. J. Burke, Numerical Electromagnetics Code—NEC-4: *Method of Moments*, Parts I (User's Manual) and II (Theory), Lawrence Livermore National Laboratory, UCRL-MA-109338, 1992.

[44] "S Parameter Design," Agilent Application Note AN-154, www.agilent.com, 2000.

Selected Bibliography

De Maio, A., M. Lops, "Design Principles of MIMO Radar Detectors," *IEEE Transactions on Aerospace and Electronic Systems,* Vol. 43, No. 3, July 2007.

Fuhrmann, D. R., G. S. Antonio, "Transmit beamforming for MIMO radar systems using partial signal correlation" *Conference Record of the Thirty-Eighth Asilomar Conference on Signals, Systems, and Computers,* Vol. 1, Nov. 7-10, 2004, pp. 295–299.

Li, J., P. Stoica, (eds.), *MIMO Radar Signal Processing,* John Wiley & Sons, Inc. 2009

Robey, F. C., et al., "MIMO radar theory and experimental results," *Conference Record of the Thirty-Eighth Asilomar Conference on Signals, Systems, and Computers,* Nov. 7–10 2004, Pacific Grove, CA.

San Antonio, G., D. R. Fuhrmann, "Beampattern synthesis for wideband MIMO radar systems," *1st IEEE International Workshop on Computational Advances in Multi-Sensor Adaptive Processing,* Dec. 13–15, 2005, pp. 105–108.

Tabrikian, J.," Barankin Bounds for Target Localization by MIMO Radars" *Fourth IEEE Workshop on Sensor Array and Multichannel Processing,* July 12–14, 2006, pp. 278–281.

Xu, L., J. Li, P. Stoica, "Adaptive Techniques for MIMO Radar," *Fourth IEEE Workshop on Sensor Array and Multichannel Processing,* July 12-14, 2006, pp. 258–262.

4

MIMO Radar Applications

This chapter is devoted to applications of the MIMO radar techniques presented in Chapter 3. We begin with airborne GMTI radar where MIMO techniques provide enhancements to systems with limited antenna aperture or spatial channels due to platform SWAP constraints. We then present an application of the same MIMO antenna to maritime radar. Next we present an application to MIMO techniques to OTH radar, where the capability of a MIMO antenna to simultaneously steer or adapt the transmit and receive spatial response in the receiver signal processor is used to mitigate the complex clutter that is colored by the complex OTH propagation channel. We conclude with a discussion of a newly emerging MIMO radar application, automotive radar, which provides a good segue into Chapters 5 and 6 and which also provides a theory for optimizing MIMO radars that was developed to support future applications including cognitive and fully adaptive radar modes that not only take advantage of the transmit spatial diversity of a MIMO system, but also perform fine-grained adaptation of the space-time waveforms on the fly based on the observed interference and target environment.

4.1 GMTI Radar Introduction

GMTI radar typically involves a pulsed-Doppler radar on an airborne platform operating in a look-down geometry. Doppler processing is used to discriminate moving ground targets from stationary ground clutter. The motion of the radar platform results in a Doppler spread of the ground clutter that for a given range bin is generally a function of azimuth. Therefore, the ability to separate targets from clutter requires a well-behaved antenna with generally narrow azimuthal mainbeam and low sidelobes in the azimuth dimension. When the sidelobes are low the discrimination of targets from clutter is limited by the width of the antenna mainbeam for a given platform speed and look direction. This is illustrated in Figure 4.1 for the side-looking case. A key GMTI parameter is the Doppler shift at which a target with prescribed signal-to-clutter ratio (SCR) can be reliably detected. This Doppler shift is called the minimum detectable velocity (MDV).

Figure 4.2 shows a table of radar parameters and a radar power budget. These parameters represent a typical radar system sized for a small UAV platform. The computed SNR indicates a detection range in excess of 70 km. Figure 4.3 shows a number of examples of simulated GMTI clutter for this set of radar parameters that

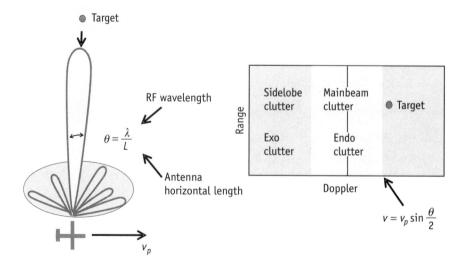

Figure 4.1   Relationship between antenna aperture size and separability of targets and ground clutter.

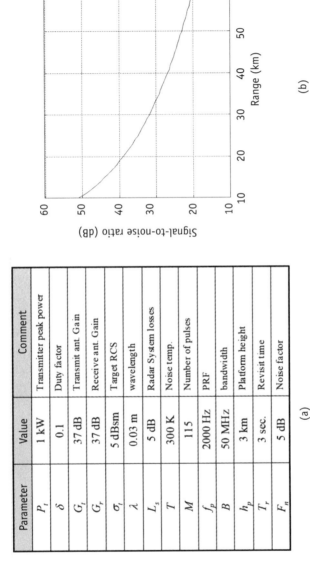

Parameter	Value	Comment
P_t	1 kW	Transmitter peak power
δ	0.1	Duty factor
G_t	37 dB	Transmit ant. Gain
G_r	37 dB	Receive ant. Gain
σ_t	5 dBsm	Target RCS
λ	0.03 m	wavelength
L_s	5 dB	Radar System losses
T	300 K	Noise temp.
M	115	Number of pulses
f_p	2000 Hz	PRF
B	50 MHz	bandwidth
h_p	3 km	Platform height
T_r	3 sec.	Revisit time
F_n	5 dB	Noise factor

(a)

Figure 4.2 (a) Typical GMTI radar system parameters for a small UAV platform. (b) SNR versus range for a single CPI (115 pulses).

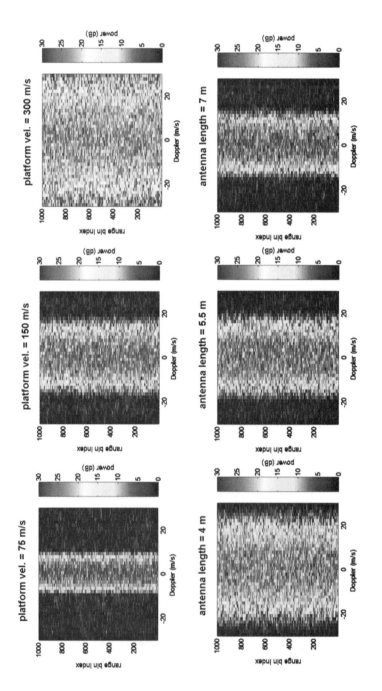

Figure 4.3 Clutter simulation examples showing the impact of platform velocity (top) and antenna aperture size (bottom) on clutter Doppler spread.

illustrate the relationship of ground clutter spread versus platform speed and antenna size. We will show below that the larger effective aperture provided by MIMO radar configurations will result in improved capability to detect slow-moving targets.

High-velocity targets that are well separated from clutter are readily detected using a *nonadaptive* and *factored* processing strategy where the radar pulses in the CPI are used to detect targets via traditional Doppler processing followed by bearing estimation employing the available system spatial channels. Most GMTI systems employ at least two antenna channels that are used to generate high accuracy bearing estimates for each target detection using advanced array processing techniques including maximum likelihood angle estimation (e.g., [1]). The bearing, range, and Doppler estimates are typically used as input to a tracking algorithm that outputs geolocation estimates and tracks for each target in the scene.

Slow-moving targets are more challenging to detect and geolocate and require more advanced system configurations and signal processing algorithms. Systems with requirements to detect targets with radial velocities that are less than the mainbeam clutter spread typically employ STAP [2]. STAP jointly combines the system temporal and spatial channels to form a space-time filter that nulls the ground clutter and preserves the target signal return. It is well known that STAP provides improved system MDV [2]; however, it places requirements on the number of system spatial channels. In order to cancel clutter and detect targets in the mainbeam clutter, at least two antenna channels are needed. In order to cancel clutter *and* compute high accuracy bearing estimates (e.g., maximum likelihood bearing estimation), at least three spatial channels are needed. We will show below that the MIMO virtual antenna elements can significantly improve system performance when the number of physical spatial channels is limited, which is often the case with small low-cost radar system designed for UAV platforms where SWAP is limited.

A comparison of the target SINR [2] for a simple two-channel airborne GMTI radar with two receive channels is shown in Figure 4.4. In each case multibin post-Doppler STAP is employed with five adjacent Doppler DoF [2]. Here we have assumed that the cross correlation between the MIMO waveforms is zero. As expected

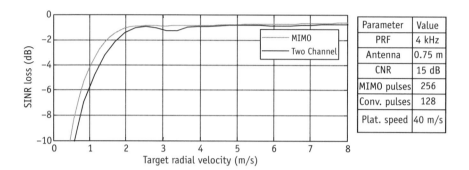

Figure 4.4 Comparison of target SINR for a MIMO and conventional GMTI radar. The two-channel curve is the conventional system. Basic system parameters are shown on the right.

the MIMO system results in a narrower clutter notch due to the high-resolution properties of the MIMO antenna arising from the extended virtual aperture. Typically, the improvement in MDV is modest; however, the MIMO system is fundamentally better for detecting slow-moving targets. A key benefit of the MIMO system is in simultaneously detecting and geolocating targets with a two-channel system, as we will show next.

After a target is detected in a GMTI system a bearing estimate is formed. A common approach is to form a maximum likelihood estimate of the target bearing. Ideally, this requires a joint two-dimensional scanning of the space-time response in azimuth and Doppler to find a peak response. In practice, it is usually sufficient to scan the space-time antenna pattern versus angle for the Doppler bin of the detected target. This is computed as [3, 4]

$$\hat{\theta} = \arg\max_{\theta} \left| \mathbf{w}'(\theta, \hat{f}) \mathbf{y}_s \right|^2 \tag{4.1}$$

where $\mathbf{w}(\theta, \hat{f})$ is the space-time filter computed according to (3.20), and $\hat{f}$ is the Doppler estimate (Doppler bin frequency) of the detected target. When the target Doppler shift is well-separated from the mainbeam clutter this estimate will be well-behaved. When the target Doppler shift is small and the target return is close to the clutter, the estimates will degrade. This is illustrated in Figure 4.5 where we show the maximum likelihood angle estimation surface for MIMO and a traditional radar for a target at 90° azimuth angle

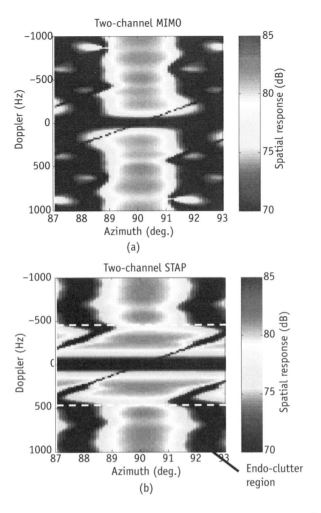

Figure 4.5 Comparison of the maximum likelihood bearing estimation surface for (a) MIMO processing and (b) conventional processing.

(boresight). In this example we are using the ideal clutter covariance matrix R_t to compute $\mathbf{w}'(\theta, \hat{f})$. As expected, the MIMO surface is much better behaved in the endo-clutter region (Doppler shifts with absolute value below 500 Hz in this example).

A cut in azimuth for the case when the target Doppler is -188 Hz is shown in Figure 4.6. The MIMO response exhibits a clear peak at the true target azimuth, whereas the conventional system has a very flat response that will lead to poor geolocation estimates. As discussed above, the MIMO system performs better because it has added effective spatial DoF to support simultaneous clutter

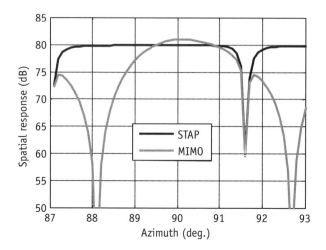

Figure 4.6   Vertical cut at Doppler = –188 Hz through surfaces shown in Figure 4.5. STAP is the traditional two-channel system.

nulling and bearing estimation. We note that if the number of physical channels is increased to three, the clear differences between MIMO and traditional system becomes much less pronounced for the endo-clutter target case; however, the MIMO system will exhibit fundamentally better performance, as we will show next.

Figure 4.7 shows the bearing estimation performance of the MIMO and traditional system as a function of target Doppler and SNR. These results were computed using 5,000 Monte Carlos trials. We see that when the target is well-separated from the clutter (high speed) the relative bearing estimation performance improvement between the MIMO and conventional arrays obeys the bound provided in (3.24) and (3.25) of √2. When the target is in the clutter (endo-clutter) the difference between the MIMO and conventional antenna is much more pronounced. For this example the improvement in bearing estimation accuracy of the MIMO antenna is greater than 2× for some values of SNR. In this case the MIMO system provides the performance of an antenna with twice the aperture!

4.2   Low-Cost GMTI MIMO Radar

In this section we present a case study on how a real-world radar system can be modified to add MIMO capabilities. The system, shown in Figure 1.3, is a low-cost radar designed to operate

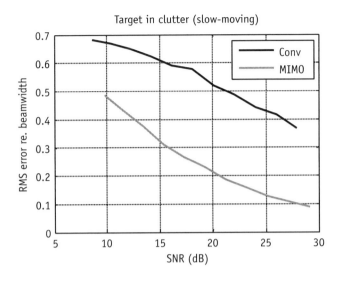

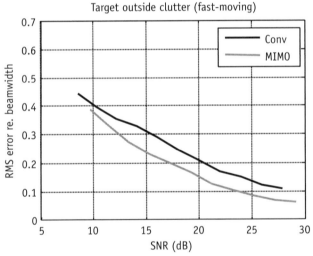

Figure 4.7 Comparison of bearing estimation performance for slow (top) and fast (bottom) moving targets. Conv is the traditional single-waveform system.

on UAV platforms and therefore has very stringent size, weight, and power constraints. The original radar involves a small antenna needed to meet platform payload constraints. The antenna has two phase centers and backend electronics to support two-channel processing. As discussed earlier in this chapter, the original system that was specifically designed for small UAV platforms and was meant to detect faster-moving targets provides a limited GMTI capability against slower-moving targets because it is limited to two channels. Thus, the system is an ideal candidate for MIMO processing. By transmitting low correlation waveforms from each of the two channels, additional spatial channels needed to detect slow-moving targets *and* compute high accuracy bearing estimates can be synthesized.

One of the key challenges with applying MIMO techniques is to find a method that allows for efficient separation of the waveforms that does not significantly impact system performance and implementation cost and complexity. As discussed in Chapter 3, in theory there are many options for waveforms that provide separability of the transmit channels. An obvious and attractive choice is the use of CDMA-type waveforms where a unique phase coded pulse is used on each channel. This type of waveform is attractive because it allows both channels to transmit in the same band and at the same time limiting the impact on radar timeline and required bandwidth. Also, these waveforms can be transmitted with a constant modulus allowing for the use of efficient power amplifiers.

The main challenge with using CDMA-type waveforms for airborne GMTI applications with strong distributed clutter is that the cross correlation between waveforms can result in significant clutter leakage among channels if the waveforms are not perfectly orthogonal, which is always the case with coded waveforms (as was shown in Chapter 3). Even though the cross-correlation noise due to a single clutter radar resolution cell leaking from one waveform into the matched filter of another waveform will be small since the waveforms can often have 30 to 40 dB of isolation, the accumulation of the noise from all the radar resolution cells can easily exceed the thermal noise floor and desensitize the system [5, 6].

This is illustrated in Figure 4.8 where we have simulated a two-channel X-band GMTI radar operating in distributed clutter with a single target present. The result is shown for a traditional single-waveform system employing an LFM waveform, a traditional system with a random phase-coded waveform, and a MIMO system with two random phase-coded waveforms. In all cases the spatial channels have been coherently steered to broadside of the antenna array. For the first two cases this is simply conventional beamforming of the receive antenna elements. For the MIMO system the receive and transmit elements are jointly combined to form the beam pattern. In all cases the resulting response will be a two-way

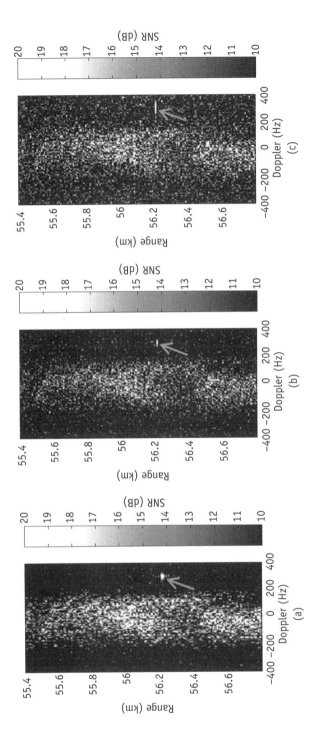

Figure 4.8 Range-Doppler clutter maps: (a) traditional radar with LFM waveforms, (b) traditional radar with random phase-coded waveform, and (c) MIMO radar with two random phase coded waveforms. The arrow points to a target in the scenario.

pattern of the radar antenna aperture as discussed in Chapter 3. The mainbeam ground clutter is clearly visible in all cases around zero Doppler. For the LFM system the clutter is isolated in Doppler. For the phase-coded waveforms the background noise is raised due to the clutter leaking through the generally higher-range sidelobes of the phase-coded waveform. We see that the background noise is clearly the highest for the MIMO system. This is due to the added cross-correlation noise between waveforms introduced by all the clutter returns in the scene. This noise will reduce the sensitivity of the radar and limit the performance improvements of the MIMO configuration as was shown in Chapter 3. We note that better waveform design, including designing CDMA waveforms with better isolation, is a potential way to reduce the impact of the clutter leakage problem. For most systems, however, the strong distributed clutter background will make this approach very difficult.

Fortunately, approaches with non-CDMA waveforms have been shown to provide good performance for MTI applications. A common approach is DDMA (e.g., [7]), where each transmit channel is formed using a unique phase code from pulse to pulse. For example, each channel might have a unique phase ramp (Doppler shift) such that the signal from each transmitter can be directly separated in the Doppler domain. This is illustrated in Figure 4.9 for a MIMO system with three waveforms. For the two-channel system under consideration here the phase ramps shown in Figure 4.10 would be applied to each antenna, resulting in the signal and clutter returns being separated in each half of the unambiguous Doppler space of the radar waveform. This approach works better than the CDMA-type waveforms because it provides better isolation between the channels. We will see below that this approach is effectively TDMA, which provides truly orthogonal waveforms. Also, since we can use any waveform envelope on an individual pulse we are free to use waveforms such as an LFM that allows the distributed clutter returns from longer ranges to be sufficiently attenuated to avoid the sensitivity losses observed in Figure 4.8.

As with any MIMO implementation, the use of unique waveforms on each channel implies a unique set of transmit electronics behind each channel. For example, to add MIMO to the existing two-channel radar would seem to require adding an additional waveform generator and power amplifier to generate a second

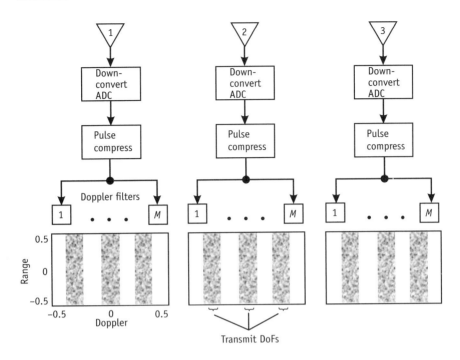

Figure 4.9 Doppler domain MIMO radar implementation.

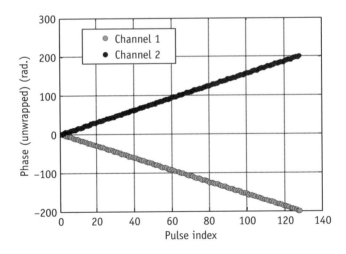

Figure 4.10 DDMA waveform phase response versus pulse index.

waveform. This is illustrated in Figure 4.11. For many systems this added hardware is impractical in terms of cost, size, weight, and power. In the case of the existing two-channel radar, it was not practical to add this additional hardware and still meet the system

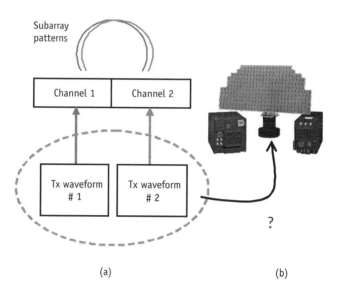

Figure 4.11 (a) Straightforward MIMO implementation with multiple waveform generators. (b) Exisiting radar with single transmit waveform that is fed to the antenna through a rotary joint feed.

cost and SWAP requirements. In fact, it is likely that it would have been more difficult to add the additional transmit hardware than to actually add additional electronics to support more physical receiver channels! As shown in Figure 4.11, as is typical with airborne radars requiring 360° coverage, the existing radar employs a mechanically rotating antenna with a rotary joint specially designed to allow the high-power RF transmitter signals to pass through to the rotating antenna. In this case, even if two waveform generators and power amplifiers could be added to the radar electronics, it would require a redesign of the rotary joint to allow two RF signals to pass through to the antenna.

We will show next that by departing from theoretical MIMO architectures that were mostly developed from a signal processing point of view, it is often possible to take advantage of MIMO without significantly impacting system cost and SWAP. In fact, some researchers have put forth compelling analysis arguing that MIMO implementations are actually more cost-effective than single-waveform systems [8]. This is especially true for AESA systems when the cost of thermal management of the hardware is included in the system design.

We begin by assuming a DDMA approach where each MIMO channel is produced by applying a unique Doppler shift (phase ramp) on each transmit channel/antenna so that the signals are uniformly spaced within the Doppler spectrum (e.g., [7]) (see Figure 4.9). The Doppler frequency applied to the nth channel is

$$f_n = \left(n - \frac{N_s - 1}{2}\right)\frac{f_p}{N_s}, n = [0, 1, \ldots, N_s - 1] \tag{4.2}$$

where N_s is the number of transmit channels and f_p is the PRF. Then the phase of the nth channel is

$$v_{n,m} = e^{j2\pi f_n T_p m} \tag{4.3}$$

where T_s is the PRI and m is the pulse index. From an antenna array point of view, the effect of the waveforms is to put a relative phase difference on each antenna input, which will modify the far field antenna pattern on each pulse. Substituting for f_n, this becomes

$$v_{n,m} = e^{j2\pi\left(n - \frac{N_s - 1}{2}\right)\frac{m}{N_s}} \tag{4.4}$$

The phase of the response, $v_{n,m}$, is tabulated in the first two columns of the table in Figure 4.12 for the first four pulses in the coherent processing interval for a two-channel system ($N_s = 2$). We see that the signals on each channel are in phase on the odd pulses and 180° out of phase on the even pulses. Thus the system will transmit with a sum pattern on the odd pulses and a difference beam on the even pulses as shown in the top plot in Figure 4.13. This can be further shown by subtracting the phase of $v_{n,m}$ for adjacent channels n and $n+1$, which gives

$$\Delta\phi = \frac{2\pi}{N_s}m \tag{4.5}$$

which for the case $N_s = 2$ is $\Delta\phi = \pi m$, where we see that the signals at the two antennas are in phase on the odd pulses and out of phase on the even pulses. It is interesting to note that for the case of

Pulse index (m)	Original phase ramps: sum and difference patterns		90° added to one channel: sequential lobing	
	$n = 0$	$n = 1$	$n = 0$	$n = 1$
0	0	0	0	$\dfrac{\pi}{2}$
1	$-\dfrac{\pi}{2}$	$\dfrac{\pi}{2}$	$-\dfrac{\pi}{2}$	π
2	$-\pi$	π	$-\pi$	$\dfrac{3\pi}{2}$
3	$-\dfrac{3\pi}{2}$	$\dfrac{3\pi}{2}$	$-\dfrac{3\pi}{2}$	0

Figure 4.12 Phase shifts on each antenna (n is the antenna index) due to the MIMO waveforms.

three MIMO waveforms the relative phasing among the channels from pulse to pulse will produce patterns that alternate between the sum pattern and full aperture beams steered forward and aft by one natural beamwidth of the antenna. Thus the MIMO system will effectively scan the antenna from pulse to pulse using natural beam patterns when the number of MIMO transmit channels is odd. When the number of channels is even some of the patterns on a given pulse may not be natural beam patterns, as was the case when $N_s = 2$ where one of the patterns was a difference pattern.

We conclude that a simple transmit antenna with a capability to switch beams pulse to pulse would produce the same transmit spatial DoF as the far more complicated DDMA waveforms. In this way, we see that the DDMA MIMO approach effectively augments a system with a monopulse on transmit capability. The transmit DoF can be separated in the receiver by simply taking every other pulse. That is, pulses 1, 3, 5,... are transmit channel one and pulses 2, 4, 6... are channel two.

As discussed in Chapter 3, the impact of transmitting MIMO waveforms on the system transmit hardware must be considered in cases when the system antenna channels have nonzero mutual coupling between transmitter antenna elements. In the case of the X-band radar being considered here, the planar antenna with

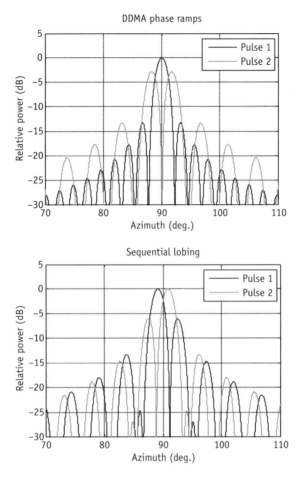

Figure 4.13 Antenna patterns resulting from DDMA waveforms. DDMA was modified to improve transmitter efficiency and VSWR

two generally large subarrays was not expected to exhibit a large amount of mutual coupling between the transmit channels since the subarrays have relatively narrow antenna patterns. It is instructive, however, to show how the VSWR of the antenna varies from pulse to pulse as the relative phasing of the antennas is varied using the DDMA approach [9].

We employ the same two-port antenna model presented in Chapter 3 for a transmitting antenna array with mutual coupling. In this case we assume an X-band antenna similar to the one being considered here; however, we assume that we have phase control on each individual antenna element and each adjacent pair of elements has mutual coupling according to the model in Chapter 3 and we assume for simplicity that all other element pairs are independent (i.e., not electrically coupled). Figure 4.14 shows the computed VSWR for the single-waveform case and the DDMA waveforms. We see that the DDMA waveforms cause a large change in VSWR from pulse to pulse for the two-channel case. In particular, the VSWR is significantly increased when the input waveforms result in the difference pattern discussed above. We note that this variation in VSWR is likely to be somewhat less pronounced for the actual X-band hardware under consideration since the model here assumes a rather high degree of coupling between the channels. However, for systems that employ individual MIMO channels with broader antenna patterns and higher levels of antenna coupling like the OTH radar examples discussed later in this chapter, this effect will be important to consider in the overall MIMO waveform and antenna design as discussed in [9].

To avoid transmitting using a difference pattern, a constant phase shift of 90° can be added to one of the DDMA waveforms as shown in the two rightmost columns in the table in Figure 4.12. This will not change the performance of the MIMO system, since the resulting two DDMA waveforms will still be independent phase ramps allowing for separation of the waveforms in the Doppler domain and the constant phase term can be accounted for in the signal processor. This results in the antenna switching between two adjacent full aperture beams from pulse to pulse (e.g., sequential lobing [10]) as opposed to the sum and difference pattern

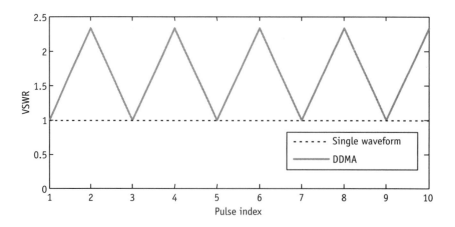

Figure 4.14 VSWR for DDMA and a traditional single-waveform system.

(monopulse on transmit). The beam patterns for the two cases are shown in the right-hand plot in Figure 4.13. We note that in this case the beams have rather high sidelobes because we are using subarrays with a phase center separation that is much larger than a half wavelength for the transmit channels.

As mentioned above, the main benefit of viewing the DDMA approach from an antenna point of view is that is allows for MIMO data to be collected with potentially less complex hardware. For example, the arbitrary waveform generators on each antenna can be replaced with much simpler phase shifters. In this case we have shown that we only need to switch the beam between two antenna positions, which can theoretically be implemented with a single phase shifter with two phase states as shown in Figure 4.15. In this case we would use a high-power phase shifter behind one of the transmitter antennas and leave the other transmitter path unmodulated. We note that instead of using two phase ramps pulse to pulse as the waveform set, this implementation uses a biphase modulated signal and constant phase signal. The key is that the

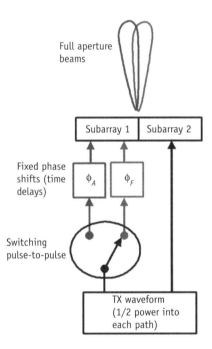

Figure 4.15 Simplified two-channel MIMO antenna architecture with the same performance as DDMA MIMO.

system imparts a temporally varying encoding of the transmit energy that can be exploited in the MIMO processor. As we have shown here, viewing a MIMO system from this perspective leads directly to simplified hardware alternatives.

It is interesting to note that any system that imparts a temporally varying encoding of the transmit energy or varying transmit spatial antenna pattern can be viewed as a MIMO system. One interesting case is the baseline scanning antenna under consideration here. In the traditional mode of operation the system collects a number of pulses for each beam position as the antenna scans. If the CPI was doubled the system would collect the same number of pulses as the corresponding MIMO system (assuming the MIMO system CPI is doubled to account for the loss in transmit gain, as discussed in Chapter 3). In this case the first half of the pulses would be for beam position one and the second half would be for beam position two. By contrast the MIMO system would use the same set of pulses; however, with the beam scanning back and forth (interleaved) pulse to pulse. Therefore, we see that the baseline scanning system and DDMA MIMO system actually collect the same data, but in a difference sequence. For targets and clutter signals that do not decorrelate in time we would expect the performance of these two systems to be similar; however, when there is significant clutter decorrelation due to effects such as internal clotter motion (ICM) the sequential lobing system will exhibit less clutter cancellation loss. As well, the sequential system has the benefit of a longer integration time on targets and will have a better capability to resolve targets in the Doppler dimension.

The sequential lobing approach to MIMO can easily produce more transmit channels by simply scanning the beam to the desired number of beam positions. For example, if three transmitter channels are desired the antenna is scanned across three adjacent beam locations from pulse to pulse. As with the traditional DDMA approach, the useable unambiguous Doppler space is reduced for each transit channel added since the PRF is effectively reduced by the number of beam locations visited during the CPI. The key advantage is that this strategy produces the MIMO channels with a very simple scanning antenna as opposed to a more complicated system employing arbitrary waveform generators behind each transmit channel.

The DDMA approach impacts the waveform ambiguity properties. The ambiguity surface for the baseline X-band radar is shown in Figure 4.16. The maximum unambiguous range is 150 km and the maximum unambiguous Doppler is 15 m/s. In the case of DDMA MIMO with the sequential beams the radar ambiguity will vary as a function of the azimuth angle to the target. For example, a target at broadside of the antenna between the two beams will have the same ambiguity as the baseline system since the target will be illuminated on every pulse with equal energy. A target at the peak of one of the sequential beams, however, will only be illuminated on every other pulse, resulting in a target return with effectively half the PRF.

This is illustrated in Figure 4.17 for three different azimuth angles. We see that the largest impact is a loss in the useable unambiguous Doppler space. For the system under consideration the loss in useable unambiguous Doppler space was acceptable since the chosen application of the new MIMO mode was to detect slow-moving targets. For cases when higher-velocity target detection and geolocation is also required, this loss may limit the utility of the MIMO approach and would need to be weighed against the overall performance benefits of MIMO. In this case an alternative would be to add a third receiver antenna phase center to achieve good slow target detection and bearing estimation and avoid the impact of MIMO on the radar ambiguity. Alternatively, if more phase control were added to each channel, it might be possible to

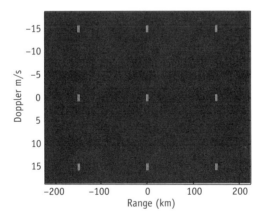

Figure 4.16 Ambiguity surface for the baseline waveform. Unambiguous Doppler is approximately 15 m/s and unambiguous range is 150 km.

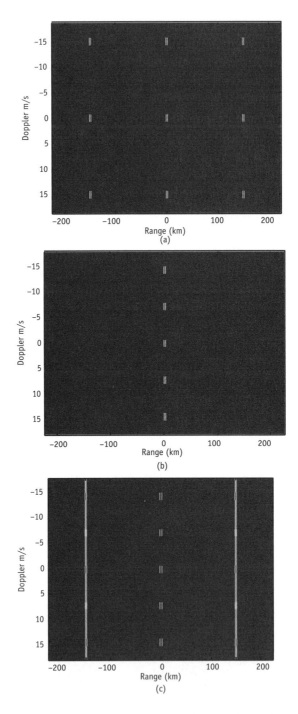

Figure 4.17   (a) MIMO ambiguity surface for signals at broadside. (b)MIMO ambiguity for signals at one beamwidth away from broadside. (c) Ambiguity surface for signal at one half beamwidth relative to broadside.

optimize the radar ambiguity to improve the unambiguous Doppler space, although this would increase complexity and cost of the hardware. Finally, depending of the system, it might be possible to operate the MIMO mode at a higher PRF (which commensurates with a reduction in pulse width to keep average transmit power the same). In this case the design would be trading usable unambiguous range for more usable unambiguous Doppler space.

The chosen system architecture is shown in Figure 4.18. It was found that using the same high-power switch behind each antenna was a better option from a calibration and channel equalization point of view. Having the same hardware in each path ensures a better overall match between the channels. In the end we see that all that is required is a high-power switch and low-complexity phase shifter at the input of each transmit antenna or subarray to produce the same transmitter DoF generated with the much more complicated DDMA waveforms. A hardware implementation of this architecture was developed using Faraday rotator phase shifter technology and integrated with the Telephonics UAV radar antenna system with two phase centers. One of the key challenges of the hardware design was to find a high-power phase shifter device

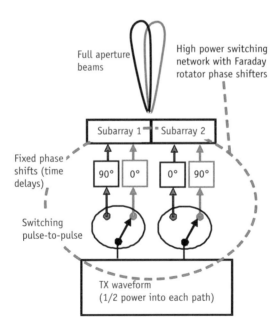

Figure 4.18 Low-cost MIMO architecture. Note that only a single-waveform generator is needed.

with low insertion loss since the phase shifter was being employed in the transmitter path. The final device used was a Faraday rotator phase shifter with two states (0° and 90°) that is switched at the system PRF as opposed to the high-power switch and constant phase shifter shown in Figure 4.18. After a number of design and test iterations a final MIMO antenna was fabricated with less the 0.5 dB of insertion loss in the transmit path. The new system was flown and used to collect data that was used to demonstrate the performance of the new low-cost MIMO mode. The experimental data processing results are described next.

The new beam-switching MIMO technique completed successful flight testing on a King Air test aircraft in May 2015. The data collections took place along the southern shore of Long Island, NY. The baseline radar consists of an 18-inch horizontal antenna aperture with two nonoverlapping receive subarrays. This antenna was originally designed to detect and track faster-moving targets but had limited bearing estimation performance for endo-clutter targets. The upgrade using the beam-switching MIMO mode demonstrated significantly improved bearing estimation performance at a fraction of the cost of adding a third receiver channel, which would have required a significant engineering effort to meet the stringent size, weight, and power requirements of this particular UAV radar system designed for small UAV platforms with limited payloads on the order of 50 lbs.

The geometry for the data collections is shown in Figure 4.19. The King Air flew up and down the southern coast of Long Island. The antenna was mechanically scanned 60° around broadside and looking out the right side of the aircraft with a revisit time at broadside of approximately 6 seconds. The scanning ensured that the radar mainbeam was continually illuminating the test area. The overland test area is also shown in Figure 4.19 and includes a moving target simulator (MTS) and a single vehicle instrumented with GPS. Data was collected with the MTS both endo and exo-clutter with an RCS of approximately 5 dBsm. The platform speed was approximately 70 m/s and the altitude was approximately 3 kft. The range to the target area was approximately 5 nm with a grazing angle of approximately 3°. Multiple passes were completed with the antenna in both the MIMO (sequential lobing) and traditional mode (fixed beam). These passes provided similar data

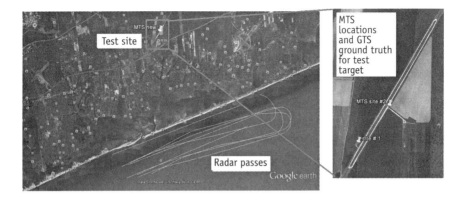

Figure 4.19 Geometry from overland data collections. The radar flew up and down the coast and looked over land to the test site, which included an MTS and a test vehicle instrumented with GPS. The white path on the zoomed in region on the right is the vehicle GPS ground truth.

sets for both MIMO and the single beam case for use in performing a side-by-side comparison. Additionally, data was collected for cases looking over water on several of the passes.

The range-Doppler clutter maps for the case when the transmit antenna is not switching are shown in Figure 4.20. These clutter maps are computed by simply Doppler processing and pulse compressing the data from a single receive antenna. In this way, they provide a good estimate of the radar transmit antenna mainbeam. The overland clutter is typically stronger and exhibits more

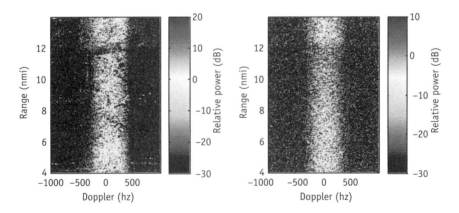

Figure 4.20 Geometry for over-land data collections. The radar flew up and down the coast and looked over land to the test site, which included an MTS and a test vehicle instrumented with GPS. The white path on the zoomed-in region on the right is the vehicle GPS ground truth.

variability due to the heterogeneous terrain and land cover. The stronger clutter observed in both cases at the longer ranges is due to wrapping which is likely an artifact of the fast convolution algorithm used to pulse compress the data.

Clutter maps for the cases when the antenna is in the MIMO mode and switching pulse to pulse are shown in Figure 4.21 for the even and odd pulses. As expected, the transmit pattern is observed to switch between a forward and aft bearing relative to the boresight of the antenna due to the MIMO switching hardware. We note that, as is typically the case with airborne MTI radar, the clutter Doppler is proportional to clutter azimuth. An estimate of the transmit beam patterns was computed by averaging the clutter in the range dimension. A comparison of the estimated beams and a

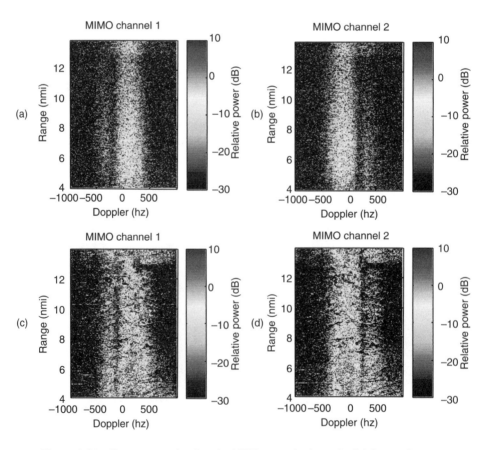

Figure 4.21 Clutter maps showing the MIMO transmit channels. (a) Even pulse over water. (b) Odd pulse over water. (c) Even pulse over land. (d) Odd pulse over land.

model of the MIMO patterns (e.g., Figure 4.13) is shown in Figure 4.22. We see that the implemented hardware produces beam patterns that closely match the desired patterns.

The two receive channels were calibrated using calibration on clutter (e.g., [11]). In this case we employed an eigen-analysis calibration-on-clutter technique. The data for each of the two receiver channels was motion-compensated to move the mainbeam clutter to zero-Doppler using the known antenna boresight and platform position and velocity. The two-channel data from the zero-Doppler bins was used to compute the 2×2 spatial covariance matrix using a large number of range bins as training data. The eigenvector associated with the largest eigenvalue of this matrix was computed. When the two channels are perfectly calibrated the amplitude and phase of the boresight signal (clutter) should be identical, and therefore the two elements of the computed eigenvector should also be identical. Any differences in the elements of this vector provide an estimate of the system spatial calibration errors.

Typically we are only interested in the relative phase errors between the channels since it is this phase error that will have the greatest impact on beamforming performance. The phase

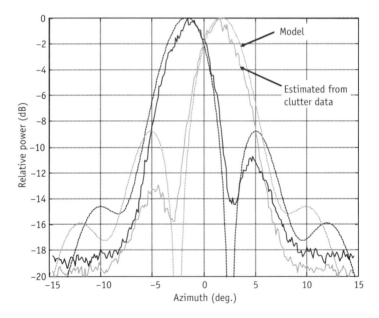

Figure 4.22 Comparison of measured and predicted MIMO transmit beam patterns. Black and gray colors indicate the two beam positions.

difference between the two channels was found to be very stable versus time with a constant error around 100°. We note that the calibration approach used here is a narrowband technique and does not account for differences in the frequency responses of the two receiver channels, which are sometimes referred to as equalization errors. For the current system where the ground clutter typically does not exceed 25-dB SNR, equalization was not found to be needed to achieve good clutter cancellation.

A similar approach was used to calibrate the transmit channels. In this case the data from a single receiver channel for both of the transmit channels (i.e., even and odd pulses) was used to form the covariance matrix used in the eigen-analysis. The calibration phase error between the two channels was also found to be relatively stable versus time, but was much smaller than the receive channel calibration errors (on the order of 5°). This happened because the transmit RF network was more mechanically fixed relative to the receive RF hardware that included a number of coaxial cable patches. Figures 4.23 and 4.24 show the full-aperture beams that are achieved when combining the two transmit channels for a single receiver channel. We see that the full aperture beam pattern generated by combining the transmit channels in the signal processor closely matches a true full aperture beam pattern that was estimated used the GMTI mode data where the transmit beam was always steered to broadside (i.e., not switching). As well, note that the high transmit pattern sidelobe observed in either of the single MIMO channel patterns (Figure 4.22) is eliminated when the two transmit channels are combined.

The calibration of the system was tested using a strong buoy target of opportunity in the scene. Figure 4.25 shows a picture of the actual navigation buoy and a typical return for a single CPI. The buoy has large corner reflectors to enhance radar returns and we see that the buoy radar return has a very high SNR. The strong stationary buoy provided a good signal to test the calibration of the system. Figure 4.26 shows the beamformed response for the buoy for three successive CPIs as the antenna mechanically scans past the buoy. The peak of this response provides the maximum likelihood estimate of the bearing to the buoy (target). The response is computed using three methods: (1) using only the receiver channels after the two transmit channels have been coherently steered

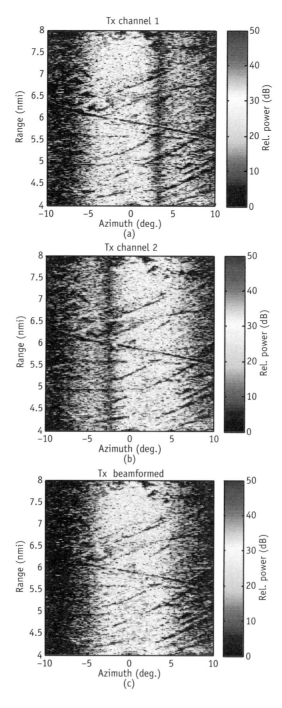

Figure 4.23 Calibration-on-clutter performance:(a) transmit channel 1, (b) transmit channel 2, and (c) two transmit channels combined to form a full aperture transmit beam.

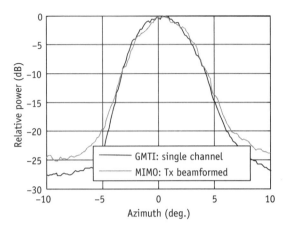

Figure 4.24 MIMO beamforming example. MIMO curve is the average versus range of the right-hand plot in Figure 4.23. The GMTI curve is computed in a similar manner using non-MIMO data.

to broadside (Rx-only), (2) using only the two transmit channels after the two receive channels have been coherently steered to broadside (Tx-only), and (3) jointly using both the receive and transmit channels (MIMO). As expected, the MIMO response is narrower and closely resembles a two-way antenna pattern, whereas the Rx-only and Tx-only responses are similar to one-way antenna patterns. This example clearly shows that the MIMO system is well calibrated and truly capable of providing a two-way antenna pattern formed in the receiver as discussed in Chapter 3!

The two transmit channels and two receive channels were combined in a MIMO-STAP beamformer to detect targets and compute bearing estimates. The overall processing flow is shown in Figure 4.27. The MIMO channels are extracted by simply reorganizing the data by pulse index (i.e., odd pulses are MIMO channel one and even pulses are MIMO channel two). This data is then motion compensated so that targets at the antenna boresight are at zero Doppler. The resulting motion-compensated data is Doppler processed and pulse compressed to complete the data preprocessing. The preprocessed data is then processed using MIMO-STAP with a multibin post Doppler STAP algorithm [2]. The data at the output of the STAP processor is thresholded to find targets. These detected targets are then passed to an MLE angle estimation algorithm where bearing estimates are computed for each detected target.

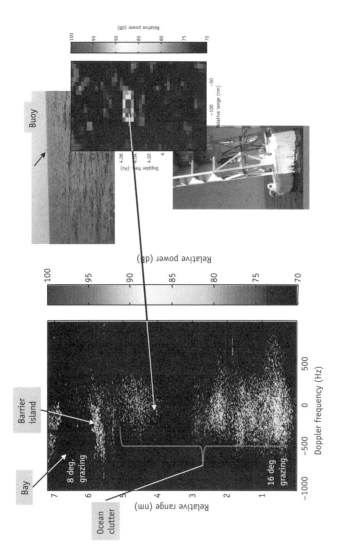

Figure 4.25 A strong buoy return was used to test the calibration of the MIMO system.

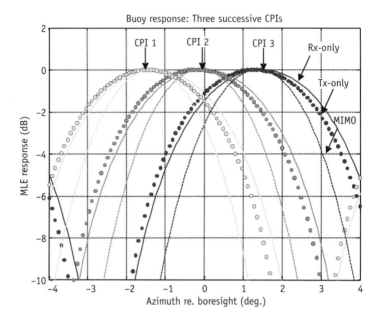

Figure 4.26 Beamformed buoy radar return.

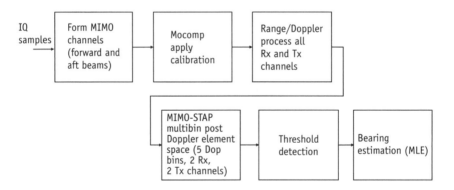

Figure 4.27 Major MIMO radar signal processing blocks.

At times during the data collection a ground truth vehicle was available with GPS; however, since this vehicle was only present in about half the data collection passes it did not provide a very large number of detections for use in developing statistics of the performance of the MIMO system. It was found, however, that the roadway in the test was rather busy with many fast-moving vehicles that were easily detected and associated with the road. An example of some of the detections with an overlay of the road translated to range-Doppler space is shown in Figure 4.28. The vehicle detections were snapped to the road, which when combined with the radar range measurement provided an accurate geolocation of the vehicle. We were able to use this information to estimate the bearing errors of the radar. The bearing error is computed by taking the difference between the target azimuth and the road azimuth at the range of the target detection. Since the range accuracy of the radar is at least an order of magnitude better than the cross-range accuracy, this approach provides a good estimate of the bearing error.

A scatter plot of all the detections along the roadway exceeding a threshold 30 dB above the thermal noise floor and with Doppler shift magnitudes greater than 300 Hz (exo-clutter) is shown in Figure 4.29. We see that the MIMO detections are clustered closer to the road due to the improved bearing estimation performance of the MIMO antenna. Plots of the estimated bearing estimation errors as a function of SNR for the exo-clutter detections along the road are shown in Figure 4.30. This plot was generated by binning the detections by estimated SNR and computing the standard deviation of the bearing errors. The plots were generated using several minutes of data resulting in several hundred detections. We see that the MIMO antenna provides improved bearing estimation performance. For the case of the conventional beamformer

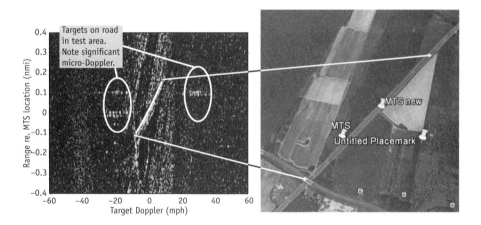

Figure 4.28 Illustration of targets of opportunity in the test scene. The highlighted road segment (white line on the left plot) was a busy roadway with many vehicles and no shadowing by trees. A number of the vehicles are circled on the plot on the left. The plot on right is a Google Earth image of the scene showing the road in relation to the MTS locations used in the collection. The radar is to the south of this scene.

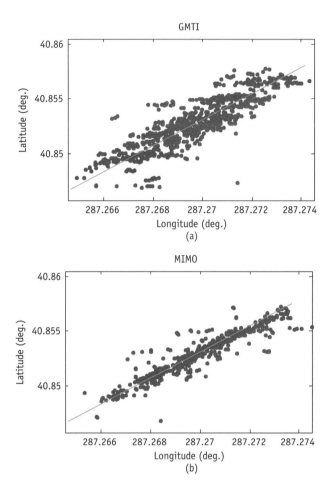

Figure 4.29 Scatter plot of exo-clutter vehicle detections on the road in the test scene. (a) Conventional GMTI radar. (b) MIMO radar. The detection threshold was set to 30 dB above the thermal noise. The detection Doppler is greater than 300 Hz. 'GMTI' is the baseline two-channel system.

the MIMO antenna provides on the order of 40% improvement in the bearing errors at the high SNR. This is consistent with the predicted performance improvement resulting from the virtual extension of the MIMO antenna aperture (Chapter 3). For the STAP cases the MIMO improvement relative to the two-channel system are slightly greater. This is likely due to the added spatial DoF of the MIMO system, which allows for more reliable simultaneous clutter mitigation and bearing estimation.

An MTS was located in the test scene. The MTS generates a radar return approximately equivalent to a 5-dBsm target. The

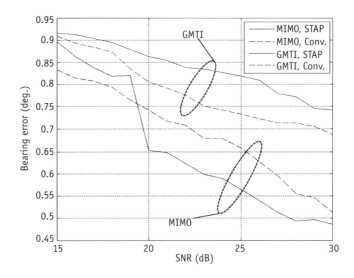

Figure 4.30 Exo-clutter bearing estimation performance comparison between the baseline two-channel systems (GMTI) and the MIMO system.

Doppler shift of the MTS was set to 94 Hz, which resulted in an endo-clutter target return. This is illustrated in Figure 4.31. After STAP processing the MTS return is clearly observed in the range-Doppler map. A large number of CPIs were collected during the flight tests and used to estimate bearing estimation errors. The results comparing the baseline two-channel system (GMTI) with the new MIMO mode (MIMO) are shown in Figure 4.32. This plot was generated by binning the detections by estimated SNR and computing the standard deviation of the bearing errors. The curves were generated using several minutes of data resulting in close to a hundred detections. We see that for the endo-clutter targets, the MIMO mode provides up to a 2× improvement in bearing estimation performance relative to the receive-only system. The relative performance improvement is greater than for the exo-clutter case shown above because, unlike the baseline receive-only system, the MIMO system provides sufficient spatial DoF to allow for both clutter mitigation and bearing estimation. We note that by adding a third receive channel this difference in performance would be reduced; however, as mentioned earlier in this chapter this would have resulted in a significant impact to system cost and SWAP, whereas the new MIMO mode employs low-cost hardware and

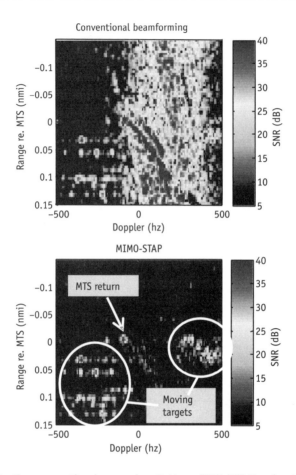

Figure 4.31   Top: conventional processing. Bottom: STAP. MTS Doppler: 97 Hz (3 kts) Radar data centered at MTS physical location MTS observed in STAP output.

was implemented in a way that had little impact on system cost and SWAP.

In summary, a new low-cost MIMO technique was presented that provides a case study on how MIMO technology can significantly improve system performance and meet challenging hardware and cost constraints. The technique was motivated by the need for additional spatial channels in a highly SWAP and cost-constrained system. It was shown that a low-cost hardware upgrade involving two high-power, low-complexity phase shifters was sufficient to provide the same MIMO capability offered by much more complex architectures involving multiple transmit channels and multiple waveform generators.

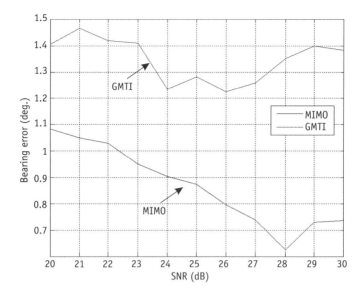

Figure 4.32 MIMO performance compared to a two-channel receiver-only GMTI mode. Bearing accuracy performance derived from detections for an endo-clutter moving target simulator.

4.3 Maritime Radar Mode

The previous section clearly showed how the MIMO antenna can improve performance for GMTI systems. The same type of antenna could be used in a maritime environment to improve detection of slow-moving vessels. In the maritime setting, most targets of interest are detected by their contrast with a generally weaker clutter background due to their very slow speeds. This is illustrated in Figure 4.33. When the clutter background is distributed, the target detectability is generally improved as the size of the radar resolution cell is reduced to the point of being the same size as the target. Thus, any mode that reduces the clutter resolution cell will typically improve detection performance. One solution is to increase the system bandwidth. If the target is sufficiently stationary, we can also increase the system integration time, which will reduce the radar Doppler resolution cell size. With a traditional system, this will require a slower scan rate and lower area coverage to maintain illumination of the target during the longer coherent processing interval. As discussed in Chapter 3, the MIMO system enables a longer integration time without sacrificing area coverage rate. Thus, the MIMO antenna can be used to spoil the illumination and

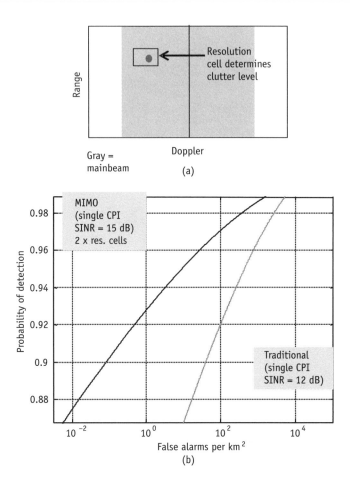

Figure 4.33 Maritime mode. (a) Maritime target detection that involves using the target contrast relative to the background clutter as opposed to target Doppler shift. (b) Detection improvement for two-channel MIMO over a traditional radar in distributed clutter with Gaussian statistics.

enable longer integration times without sacrificing two-way antenna sidelobes. As well, the MIMO system provides the improved bearing estimation accuracy discussed earlier in this chapter and in Chapter 3.

For the two-channel MIMO antenna discussed in the previous section, the MIMO radar would provide 3 dB of additional target-to-clutter ratio (contrast) relative to a traditional single-waveform system assuming noiselike distributed clutter (typically the case for lower-resolution maritime wide-area search radars). The improvement in detection performance for a 12-dB target-to-clutter ratio signal for MIMO over the traditional system is shown in Figure

4.33. Here we have assumed Gaussian clutter statistics and used the expression for pd versus pfa on a single CPI given in Chapter 3. We see that the 3-dB improvement in contrast can result in a significantly improved receiver operating characteristic (ROC) curve. We note that for scenarios when the clutter is non-Gaussian and more spiky in nature that the improvement between MIMO and the traditional system may not be as pronounced.

4.4 OTH Radar

OTH radar [12] is a technique that uses the ionospheric propagation channel to illuminate and detect targets at very long ranges. The signal propagation channel involves bouncing signals off the various layers of the ionosphere at different altitudes above the earth to enable illumination and signal returns far beyond line of sight of the radar. This mode of transmission is very effective for long-range target detection; however, it results in a very complex propagation channel that typically results in multiple modes of propagation for surface clutter, as illustrated in Figure 4.34. Additionally, the paths through the ionosphere often experience significant temporal variations, which can result in Doppler spread clutter returns with Doppler frequency extent that is similar to the Doppler shift of true target returns, making detection very difficult [12, 13].

The complex propagation channel often results in multipath clutter where the transmit angle of illumination of a radar return is different than the receive angle [14]. That is, the forward path and return paths of the clutter are different. In cases when the target and clutter returns travel different propagation paths, they will

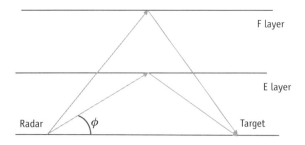

Figure 4.34 Simple illustration of the multipath OTH radar propagation environment.

generally exhibit unique combinations of launching and receiving elevation angles (e.g., [12]). In these cases it is possible to place a transmit null on the clutter path without impacting the target signal, resulting in improved target detection performance. Unfortunately the launch and receive angles of the clutter and targets are not always know a priori. Further, the clutter spatial response varies as a function of range since the propagation modes traveled are highly dependent on radar to target geometry and operating frequency. Thus, placing a single null on transmit is not be very effective for improving performance for at all ranges.

By employing a MIMO architecture, the transmit null can be adjusted as a function of range in the receive signal processor to improve detection performance at all ranges. The processing generally involves adaptive clutter cancellation of the available transmit and receive spatial DoF [8]. MIMO clutter mitigation has been demonstrated via OTH experiments and shown to result in significant detection performance. Experimental results can be found in [13, 15, 16].

4.5   Automotive Radar

Automotive radar is a growing market for both safety and autonomous vehicles. The advantages of radar relative to other potential sensors such as cameras, sonar, and lidar are the same as for many other applications: day/night and all-weather (for the most part) operation. Surprisingly, another advantage is increasingly becoming cost. The development of radar-on-a-chip technologies [17] coupled with MIMO techniques (described below) have made automotive radar solutions orders-of-magnitude lower cost than traditional radar systems.

The left-hand picture in Figure 4.35 illustrates a forward-looking automotive radar application. While range (and possibly Doppler) to an object in the field of view (FOV) is relatively straightforward to extract, angle (potentially azimuth and elevation) requires some form of integrated angle measurement. One technique is a mechanically scanned narrowbeam antenna. For a multitude of obvious and not so obvious reasons, this approach is not favored by the auto industry (or radar industry for that matter).

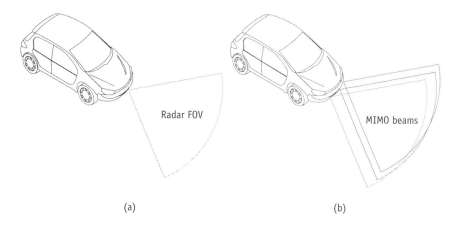

(a)   (b)

Figure 4.35   (a) Forward-looking automotive radar, and (b) automotive radar using
MIMO radar.

Phased arrays and/or electronically scanned antennas (ESAs) are a more elegant and often lower SWAP solution, but can be quite expensive relative to mechanically scanned systems. A notable exception is the metamaterial ESA developed by Echodyne (www.echodyne.com) that achieves electronic scan capability without using phase shifters, time delay units (TDUs), or multiple RF channels [18]. In addition to obtaining an accurate angle measurement, it is also desirable to minimize the number of separate radar systems required to provide adequate coverage (front, rear, and possibly sides).

Recently, MIMO techniques have garnered a great deal of attention as a means of obtaining an affordable yet effective solution for automotive radar [19]. Figure 4.18(b) illustrates the basic concept. Several small transmit apertures (and thus widebeam) are arranged to provide overlapping simultaneous coverage of a region of interest. Orthogonality could, for example, be achieved using CDMA or ODFM. A centralized receiver simultaneously receives all transmit codes and matches filters each to recover the spatial DoF. The synthetic spatial DoF can then be coherently combined in the receiver to form narrowbeams, thus creating a synthetic ESA without analog phase shifters and/or separate RF channels. Of course, as discussed in Chapter 3, MIMO techniques come at the price of lower SNR.

References

[1] Van Trees, H. L., *Optimum Array Processing: Part IV of Detection, Estimation, and Modulation Theory*, New York: John Wiley and Sons, 2002.

[2] Guerci, J. R., *Space-Time Adaptive Processing for Radar*, Second Edition, Norwood, MA, Artech House, 2014.

[3] Richmond, C. D., "Mean squared Error Threshold Prediction of Adaptive Maximum Likelihood Techniques," *Record of the Thirty-Seventh Asilomar Conference on Signals, Systems and Computers*, Monterrey, CA, November 9–12, 2003.

[4] Showman, G. A., W. L. Melvin, and D. J. Zywicki, "Application of the Cramer-Rao Lower Bound for Bearing Estimation to STAP Performance Studies," *Proceedings of the 2004 IEEE Radar Conference*, Philadelphia, April 26–29, 2004.

[5] Sun, Y., Z. He, H. Liu, and J. Li, "Airborne MIMO Radar Clutter Rank Estimation," *2011 IEEE CIE International Conference on Radar*, Chengdu, China, October 27, 2011.

[6] Forsythe, K., and D. Bliss, "MIMO Radar Waveform Constraints for GMTI," *IEEE Journal on Selected Topics in Signal Processing*, Vol. 4, No. 1, 2010.

[7] Mecca, V. F., J. L. Krolik, F. C. Robey, and D. Ramakrishnan, *Slow-Time MIMO Space Time Adaptive Processing*, ICASSP 2008, pp. 283–231.

[8] Rabideau, D. J., "MIMO Radar Aperture Optimization," MIT Lincoln Laboratory Technical Report 1149, January 25, 2011.

[9] Mecca, V., and J. Krolik, "Slow-Time MIMO STAP with Improved Power Efficiency," *2007 Conference Record of the Forty-First Asilomar Conference on Signals, Systems, and Computers*, Pacific Grove, CA, November 2007.

[10] Lo, K. W., "Theoretical Analysis of the Sequential Lobing Technique," *IEEE Transactions on Aerospace and Electronic Systems*, Vol. 35, No. 1, January 1999.

[11] Brown, M., M. Mirkin, and D. Rabideau, "Phased Array Antenna Calibration Using Airborne Radar Clutter and MIMO Signals," *48th Asilomar Conference on Signals, Systems, and Computers*, Pacific Grove, CA, November 2014.

[12] Fabrizio, G. A., *High Frequency Over-the-Horizon Radar: Fundamental Principles, Signal Processing, and Practical Applications*, New York: McGraw Hill, 2013.

[13] Frazer, G., Y. Abramovich, and B. Johnson, "HF Skywave MIMO Radar: the HiLoW Experimental Program," *Proceedings of the 2008 Asilomar Conference on Signals, Systems, and Computers*, Pacific Grove, CA, November 2008.

[14] Mecca, V. F., D. Ramakrishnan, and J. L. Krolik, "MIMO Radar Space-Time Adaptive Processing for Multipath Clutter Mitigation" *IEEE Workshop on Sensor Array and Multichannel Signal Processing*, Waltham, MA, July 2006.

[15] Abramovich, Y. I., G. J. Frazer, and B. A. Johnson, "Transmit and Receive Antenna Array Geometry Design for Spread-Clutter Mitigation in HF OTH MIMO Radar," *Proceedings of the International Radar Symposium*, Hamburg, Germany, September 9–11, 2009.

[16] Frazer, G., "Application of MIMO Radar Techniques to Over-the-Horizon Radar," *2016 IEEE International Symposium on IEEE Phased Array Systems and Technology (PAST)*, Waltham, MA, October 2016.

[17] Singh, J., B. Ginsburg, S. Rao, and K. Ramasubramanian, "AWR1642 mm Wave Sensor: 76–81-GHz Radar-on-Chip for Short-Range Radar Applications," Texas Instruments, 2017.

[18] Guerci, J. R., T. Driscoll, R. Hannigan, S. Ebad, C. Tegreene, and D. E. Smith, "Next Generation Affordable Smart Antennas," *Microwave Journal,* Vol. 57, 2014.

[19] Feger, R., C. Wagner, S. Schuster, S. Scheiblhofer, H. Jager, and A. Stelzer, "A 77-GHz FMCW MIMO Radar Based on an SiGe Single-Chip Transceiver," *IEEE Transactions on Microwave Theory and Techniques,* Vol. 57, 2009, pp. 1020–1035.

Selected Bibliography

Abramovich,Y., and Abramovich, G. Frazer, "Theoretical Assessment of MIMO Radar Performance in the Presence of Discrete and Distributed Clutter Signals," *42nd Asilomar Conference on Signals, Systems, and Computers,* Pacific Grove, CA, November 2008.

Bilik, I., et. al. "Automotive MIMO Radar for Urban Environments," *2016 IEEE Radar Conference*, Philadelphia, May 2016.

Bliss, D., and Bliss, K., Forsythe, "MIMO Radar Medical Imaging: Self-Interference Mitigation for Breast Tumor Detection," *2006 Conference Record—Asilomar Conference on Signals, Systems and Computers*, Pacific Grove, CA, 2016.

Frazer, G., J., Frazer, Y. I. Abramovich , and B. A. Johnson, "Mode-Selective MIMO OTH Radar: Demonstration of Transmit Mode-Selectivity on a One-Way Skywave Propagation Path," *2011 IEEE Radar Conference*, Kansas City, MO, May 2011.

Frazer, G. J., Y. I. Abramovich, B. A. Johnson, and F. C. Robey, "Recent Results in MIMO Over-the-Horizon Radar," *2008 IEEE Radar Conference*, Rome, Italy, May 2008.

Kantor, J., and D. Bliss, "Clutter Covariance Matrices for GMTI MIMO Radar," *2010 Conference Record—Asilomar Conference on Signals, Systems, and Computers*, Pacific Grove, CA, 2010.

Kantor, J., and D. Bliss, "Clutter Covariance Matrices for GMTI MIMO Radar," in *2010 Conference Record of the Forty Fourth Asilomar Conference on Signals, Systems, and Computers (ASILOMAR)*, Pacific Grove, CA, November 2010.

Kantor, J., and S. Davis, "Airborne GMTI Using MIMO Techniques," *2010 IEEE Radar Conference,* May 2010, pp. 1344–1349.

Li, J., et. al. "Range Compression and Waveform Optimization for MIMO Radar: A Cramér-Rao Bound Based Study, *IEEE Transactions on Signal Processing*. Vol. 56, No. 1, January 2006.

Vasanelli, C., R. Vasanelli, R. Batra, and C. Waldschmidt, "Optimization of a MIMO Radar Antenna System for Automotive Applications," *2017 11th European Conference on Antennas and Propagation (EUCAP)*, Paris, March 2017.

5

Introduction to Optimum MIMO Radar

In this chapter, we introduce the concept of optimum MIMO radar. The original papers published on MIMO radar were based on MIMO communications, in which orthogonality played a central role. However, in general, leveraging multiple inputs and outputs does not always require orthogonality. Indeed, in this chapter we will derive the optimum input-output configuration as a function of the radar channel.

In Section 5.1, we derive the optimum MIMO transmit-receive configuration for maximizing SINR, and hence target detection under usual stochastic assumptions. The optimum multi-input (MI) transmit function is shown to be the solution of an eigensystem, the kernel for which is a second-order function of the channel Green's function. While the optimum multioutput (MO) receive function is the usual Wiener-Hopf filter (colored noise matched filter, tuned to the echo of the optimum transmit function). An application to waveform optimization in the presence of colored noise is then considered.

In Section 5.2, we extend the above to the case when clutter (reverb) is present, while in Section 5.3 we apply the theory of optimum MIMO to the target ID problem, where it is shown that the optimum MI transmit function is a solution to a particular eigensystem involving second-order functions of the potential target's Green's functions.

5.1 Theory of Optimum MIMO Radar for Detection

Consider the basic radar block diagram in Figure 5.1. A generally complex-valued and multidimensional transmit signal, $\mathbf{s} \in \mathbb{C}^N$ (i.e., an N-dimensional MI signal), interacts with a target denoted by the target transfer matrix (Green's function [1]) $H_T \in \mathbb{C}^{M \times N}$. The resulting M-dimensional MO signal (echo), $\mathbf{y} \in \mathbb{C}^M$, is then received along with additive colored noise (ACN) $\mathbf{n} \in \mathbb{C}^M$ [2]. The vector-matrix formulation is completely general inasmuch as any combination of spatial and temporal dimensions can be represented.

For example, the N-dimensional input vector s could represent the N complex (i.e., in-phase and quadrature [I&Q]) [3] samples of a single-channel transmit waveform $s(t)$; that is,

$$\mathbf{s} = \begin{bmatrix} s(\tau_1) \\ s(\tau_2) \\ \vdots \\ s(\tau_N) \end{bmatrix} \tag{5.1}$$

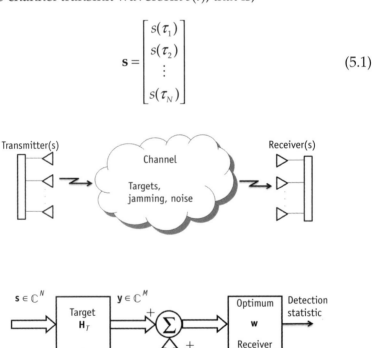

Figure 5.1 Fundamental MIMO radar block diagram for the ACN case. (After: [4].)

The corresponding target transfer matrix H_T would thus contain the corresponding samples of the complex target impulse response $h_T(t)$, which for the causal linear time invariant (LTI) case would have the form [5]

$$H_T = \begin{bmatrix} h[0] & 0 & 0 & \cdots & 0 \\ h[1] & h[0] & 0 & \cdots & 0 \\ h[2] & h[1] & h[0] & \cdots & 0 \\ \vdots & & & \ddots & \vdots \\ h[N-1] & & & h[1] & h[0] \end{bmatrix} \tag{5.2}$$

where, without loss of generality, we have assumed uniform time sampling (i.e., $\tau k = (k-1)T$), where T is a suitably chosen sampling interval [6]. Note also that without loss of generality, we have, for both convenience and a significant reduction in mathematical nomenclature overhead, chosen $N = M$; that is, the same number of transmit and receive DoF (time, space, etc.). The reader is encouraged to, where desired, reinstate the inequality and confirm that the underlying equations derived throughout this chapter have the same basic form except for differing vector and matrix dimensionalities. Note also that in general H_T is stochastic (this will be addressed shortly).

The formalism is readily extensible to the multiple transmitter, multiple receiver case. For example, if there are three (3) independent transmit and receive channels (e.g., an active electronically scanned antenna (AESA) with separate waveform generators for each transmit element or subarray), then the input vector **s** of Figure 5.1 would have the form

$$\mathbf{s} = \begin{bmatrix} \mathbf{s}_1 \\ \mathbf{s}_2 \\ \mathbf{s}_3 \end{bmatrix} \in \mathbb{C}^{3N} \tag{5.3}$$

where $\mathbf{s}_i \in \mathbb{C}^N$ denotes the samples (as in [5.1]) of the transmitted waveform out of the ith transmit channel. The corresponding target transfer matrix would in general have the form

$$H_T = \begin{bmatrix} H_{11} & H_{12} & H_{13} \\ H_{21} & H_{22} & H_{23} \\ H_{31} & H_{32} & H_{33} \end{bmatrix} \in \mathbb{C}^{3N \times 3N} \tag{5.4}$$

where the submatrix $H_{i,j} \in \mathbb{C}^{N \times N}$ is the transfer matrix between the ith receive and jth transmit channels for all time samples of the waveform.

These examples make clear that the matrix-vector, input-output formalism is completely general and can accommodate whatever transmit-receive DoF desired. Returning to Figure 5.1, we now wish to jointly optimize the transmit and receive functions. We will find it convenient to work backward; that is, begin by optimizing the receiver as a function of the input, then finally optimize the input—and thus the overall output SINR.

For any finite norm input **s**, the receiver that maximizes output SINR for the ACN case is the so-called whitening (or colored noise) matched filter, as shown in Figure 5.2 [2]. Note that for the additive Gaussian colored noise (AGCN) case, this receiver is also statistically optimum [2].

If $R \in \mathbb{C}^{N \times N}$ denotes the total interference covariance matrix associated with **n**, which is further assumed to be independent of **s** and Hermitian positive definite [7] (guaranteed in practice due to ever-present receiver noise [2]), then the corresponding whitening filter is given by [2]

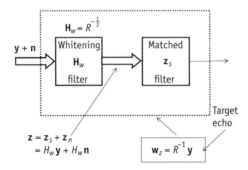

Figure 5.2 The optimum receiver for the ACN case consists of a whitening filter followed by a white noise matched filter. (After: [4].)

$$H_w = R^{-\frac{1}{2}}$$

(5.5)

The output of the linear whitening filter $\mathbf{z} \in \mathbb{C}^N$ will consist of signal and noise components, $\mathbf{z}_s$, $\mathbf{z}_n$, respectively, given by

$$
\begin{aligned}
\mathbf{z} &= \mathbf{z}_s + \mathbf{z}_n \\
&= H_w \mathbf{y}_s + H_w \mathbf{n} \\
&= H_w H_T \mathbf{s} + H_w \mathbf{n}
\end{aligned}
$$

(5.6)

where $\mathbf{y}_s \in \mathbb{C}^N$ denotes the target echo as shown in Figure 5.2 (i.e., the output of H_T).

Since the noise has been whitened via a linear—in this case full rank—transformation [2], the final receiver stage consists of a white noise matched filter of the form (to within a multiplicative scalar)

$$\mathbf{w}_z = \mathbf{z}_s \in \mathbb{C}^N$$

(5.7)

The corresponding output SNR is thus given by

$$
\begin{aligned}
\mathrm{SNR}_o &= \frac{\left| \mathbf{w}_z' \mathbf{z}_s \right|^2}{\mathrm{var}\left(\mathbf{w}_z' \mathbf{z}_n \right)} \\
&= \frac{\left| \mathbf{z}_s' \mathbf{z}_s \right|^2}{\mathrm{var}\left(\mathbf{z}_s' \mathbf{z}_n \right)} \\
&= \frac{\left| \mathbf{z}_s' \mathbf{z}_s \right|^2}{E\left\{ \mathbf{z}_s' \mathbf{z}_n \mathbf{z}_n' \mathbf{z}_s \right\}} \\
&= \frac{\left| \mathbf{z}_s' \mathbf{z}_s \right|^2}{\mathbf{z}_s' E\left\{ \mathbf{z}_n \mathbf{z}_n' \right\} \mathbf{z}_s} \\
&= \frac{\left| \mathbf{z}_s' \mathbf{z}_s \right|^2}{\mathbf{z}_s' \mathbf{z}_s} \\
&= \left| \mathbf{z}_s' \mathbf{z}_s \right|
\end{aligned}
$$

(5.8)

where var(·) denotes the variance operator. Note that due to the whitening operation $E\left\{ \mathbf{z}_n \mathbf{z}_n' \right\} = I$.

In other words, the output SNR is proportional to the energy in the whitened target echo. This fact is key to optimizing the input function: Chose $\mathbf{s}$ (the input) to maximize the energy in the whitened target echo; that is,

$$\max_{\{\mathbf{s}\}} \left| \mathbf{z}_s' \mathbf{z}_s \right| \tag{5.9}$$

Substituting $\mathbf{z}_s = H_w H_T \mathbf{s}$ into (5.9) yields the objective function that explicitly depends on the input

$$\max_{\{\mathbf{s}\}} \left| \mathbf{s}'(H'H)\mathbf{s} \right| \tag{5.10}$$

where

$$H \triangleq H_w H_T \tag{5.11}$$

Recognizing that (5.10) involves the magnitude of the inner product of two vectors $\mathbf{s}$ and $(H'H)\mathbf{s}$, we readily have from the Cauchy-Schwarz theorem [8], the condition in which $\mathbf{s}$ must satisfy to yield a maximum, namely $\mathbf{s}$ *must be collinear with* $(H'H)\mathbf{s}$; that is,

$$(H'H)\mathbf{s}_{opt} = \lambda_{max}\mathbf{s}_{opt} \tag{5.12}$$

In other words, the optimum input $\mathbf{s}_{opt}$ must be an eigenfunction of $(H'H)$ with associated maximum eigenvalue.

It is important to recognize that the above set of input-output design equations represent the absolute optimum that any combination of transmit-receive operations can achieve. Thus, they are of fundamental value to the radar systems engineer interested in ascertaining the value of advanced adaptive methods (e.g., adaptive waveforms, transmit-receive beamforming). Note also that (5.12) can be generalized to the case where the target response is random; that is,

$$E(H'H)\mathbf{s}_{opt} = \lambda_{max}\mathbf{s}_{opt} \tag{5.13}$$

In this case, $E(\ )$ denotes the expectation operator, and s_{opt} maximizes the expected value of the energy in the whitened target echo.

Next, we illustrate the application of the above optimum design equations to the additive colored noise problem arising from a broadband multipath interference source.

This example illustrates the optimum transmit-receive configuration for maximizing output SINR in the presence of colored noise interference arising from a multipath broadband noise source. More specifically, for the single transmit-receive channel case, it derives the optimum transmit pulse modulation (i.e., pulse shape).

Figure 5.3 illustrates the situation at hand. A nominally broadband white noise source undergoes a series of multipath scatterings that in turn colors the noise spectrum [9]. Assuming (for simplicity) that the multipath reflections are dominated by several discrete specular reflections, the resultant signal can be viewed as the output of a causal tapped delay line filter (i.e., an FIR filter [5]) of the form

$$h_{mp}[k] = \alpha_0 \delta[k] + \alpha_1 \delta[k-1] + \ldots + \alpha_{q-1}\delta[k-q-1] \qquad (5.14)$$

that is driven by white noise. The corresponding input-output transfer $H_{mp} \in \mathbb{C}^{N \times N}$ is thus given by

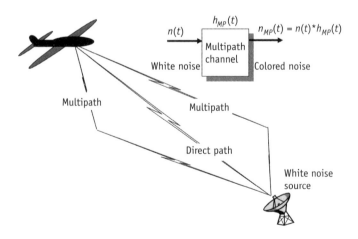

Figure 5.3   Illustration of colored noise interference resulting from a broadband (i.e., white noise) source undergoing multipath reflections. (After: [4].)

$$H_{mp} = \begin{bmatrix} h_{mp}[0] & 0 & \cdots & 0 \\ h_{mp}[1] & h_{mp}[0] & & \vdots \\ \vdots & & \ddots & 0 \\ h_{mp}[N-1] & \cdots & h_{mp}[1] & h_{mp}[0] \end{bmatrix} \qquad (5.15)$$

In terms of the multipath transfer matrix, H_{mp}, the colored noise interference covariance matrix is given by

$$\begin{aligned} E(\mathbf{nn'}) &= E\left(H_{mp}\mathbf{vv}H'_{mp}\right) \\ &= H_{mp}E(\mathbf{vv})H'_{mp} \\ &= H_{mp}H'_{mp} \\ &= R \end{aligned} \qquad (5.16)$$

where the driving white noise source $v \in \mathbb{C}^N$ is a zero mean complex vector random variable with an identity covariance matrix; that is,

$$E\{\mathbf{vv'}\} = I \qquad (5.17)$$

Assuming a unity gain point target at the origin (i.e., $h_T[k] = \delta[k]$) yields a target transfer matrix $H_T \in \mathbb{C}^{N \times N}$ given by

$$\begin{aligned} H_T &= \begin{bmatrix} h_T[0] & 0 & \cdots & 0 \\ h_T[1] & h_T[0] & & \vdots \\ \vdots & & \ddots & 0 \\ h_T[N-1] & \cdots & h_T[1] & h_T[0] \end{bmatrix} \\ &= I \end{aligned} \qquad (5.18)$$

While certainly a more complex (and thus realistic) target model could be assumed, we wish to focus on the impact the colored noise has on shaping the optimum transmit pulse. We will introduce more complex target response models in the target ID section.

Figure 5.4 shows the in-band interference spectrum for the case when $\alpha_0 = 1$, $\alpha_2 = 0.9$, $\alpha_5 = 0.5$, $\alpha_{10} = 0.2$, all other coefficients are set

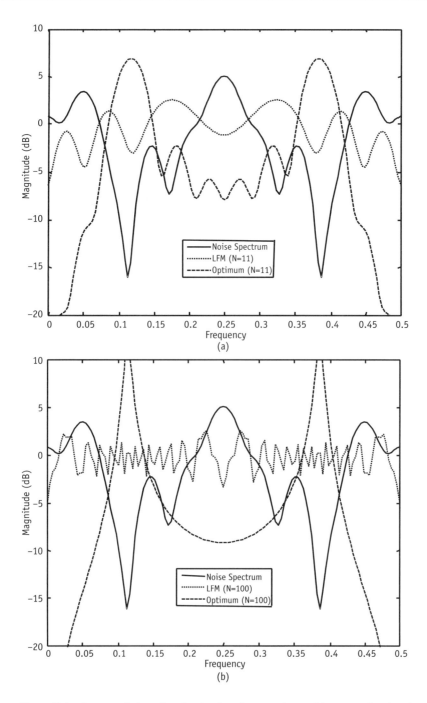

Figure 5.4 Spectra of the colored noise interference along with conventional and optimal pulse modulations: (a) short pulse case, and (b) long pulse case. Note that in both cases the optimum pulse attempts to antimatch to the colored noise spectrum under the frequency resolution constraint set by the total pulse width.

to zero. The total number of fast time (range bin) samples was set to both a short pulse case of (Figure 5.4[a]), and a long pulse case of (Figure 5.4[b]). Note that the multipath colors the otherwise flat noise spectrum. Also displayed is the spectrum of a conventional (and thus nonoptimized) LFM pulse with a time-bandwidth product $(\beta\tau)$ of 5 (Figure 5.4[a]) and 50 (Figure 5.4[b]), respectively [10, 11].

Given R from (5.16), the corresponding whitening filter H_w is given by

$$H_w = R^{-\frac{1}{2}} \tag{5.19}$$

Along with (5.18), the total composite channel transfer matrix H is thus given by

$$
\begin{aligned}
H &= H_w H_T \\
&= H_w \\
&= R^{-\frac{1}{2}}
\end{aligned}
\tag{5.20}
$$

Substituting (5.20) into (5.12) yields

$$R^{-1}\mathbf{s}_{opt} = \lambda \mathbf{s}_{opt} \tag{5.21}$$

That is, the optimum transmit waveform is the maximum eigenfunction associated with the inverse of the interference covariance matrix. The reader should verify that this is also the minimum eigenfunction of the original covariance matrix R—and thus can be computed without matrix inversion.

Displayed in Figures 5.4(a) and (b) are the spectra of the optimum transmit pulses obtained by solving (5.21) for the maximum eigenfunction/eigenvalue pair for the aforementioned short and long pulse cases, respectively. Note how the optimum transmit spectrum naturally emphasizes those portions of the spectrum where the interference is weak—an intuitively satisfying result.

The SINR gain of the optimum short pulse, SINR$_{opt}$, relative to that of a nonoptimized chirp pulse, SINR$_{LFM}$, is

$$\text{SINR}_{\text{gain}} \overset{\Delta}{=} \frac{\text{SINR}_{\text{opt}}}{\text{SINR}_{\text{LFM}}} = 7.0 \text{ dB} \tag{5.22}$$

while for the long pulse case

$$\text{SINR}_{\text{gain}} \overset{\Delta}{=} \frac{\text{SINR}_{\text{opt}}}{\text{SINR}_{\text{LFM}}} = 24.1 \text{ dB} \tag{5.23}$$

The increase in SINR for the long pulse case is to be expected since it has finer spectral resolution, and can therefore more precisely shape the transmit modulation to antimatch the interference. Of course, the unconstrained optimum pulse has certain practical deficiencies (such as poorer resolution and compression sidelobes) compared to a conventional pulse. This can be addressed by performing constrained optimization (see [4] for examples).

The above example is similar in spirit to the spectrum notching waveform design problem that arises when there are strong cochannel narrowband interferers present [12]. In this case it is not only desirable to filter out the interference on receive, but also to choose a transmit waveform that minimizes energy in the cochannel bands. The reader is encouraged to experiment with different notched spectra and pulse length assumptions and applying (5.12). Nonimpulsive target models (i.e., extended targets) can also be readily incorporated.

5.2 Optimum MIMO Radar in Clutter

Consider Figure 5.5, which depicts the target and clutter channels. By definition, clutter is unwanted target returns, such as ground reflections in MTI radar.

Unlike the previous colored noise case in Section 5.1, clutter is a form of signal-dependent noise [13, 14]—since the clutter returns depend on the transmit signal characteristics (e.g., transmit antenna pattern and strength, operating frequencies, bandwidths, polarization). Referring to Figure 5.5, the corresponding SCNR at the input to the receiver is given by

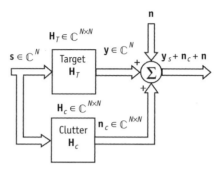

Figure 5.5   Radar signal block diagram for the clutter dominant case illustrating the direct dependency of the clutter signal on the transmitted signal.

$$
\begin{aligned}
\text{SCR} &= \frac{E\left(\mathbf{y}_T'\mathbf{y}_T\right)}{E\left(\mathbf{y}_c'\mathbf{y}_c\right)+\sigma^2 I} \\[2mm]
&= \frac{\mathbf{s}'E\left(H_T'H_T\right)\mathbf{s}}{\mathbf{s}'E\left(\left(H_c'H_c\right)+\sigma^2 I\right)\mathbf{s}}
\end{aligned}
\tag{5.24}
$$

where $H_c \in C^{N \times N}$ denotes the clutter transfer matrix that is generally stochastic, and $\sigma^2 I$ denotes the covariance of the white noise (nonclutter) component. Equation (5.24) is a generalized Rayleigh quotient [7] that is maximized when $\mathbf{s}$ is a solution to the generalized eigenvalue problem

$$
E\left(H_T'H_T\right)\mathbf{s} = \lambda\left(E\left(H_c'H_c\right)+\sigma^2 I\right)\mathbf{s}
\tag{5.25}
$$

with corresponding maximum eigenvalue. Since $(E(H'_cH_c)+\sigma^2 I)$ is positive definite, (5.25) can be converted to an ordinary eigenvalue problem of the form we have already encountered, specifically,

$$
\left(E\left(H_c'H_c\right)+\sigma^2 I\right)^{-1} E\left(H_T'H_T\right)\mathbf{s} = \lambda\mathbf{s}
\tag{5.26}
$$

The application of (5.25)–(5.26) to the full-up space-time clutter suppression of MTI clutter is available in [4]. Here we will consider its application to the sidelobe target suppression problem—which of course is at the heart of the ground clutter interference issue since clutter is essentially unwanted target returns.

Consider a narrowband $N = 16$ element ULA with half-wavelength interelement spacing and a quiescent pattern displayed in

Figure 5.6. This is essentially the ULA we have previously considered earlier in this book (e.g., Chapter 2). In addition to the desired target at a normalized angle of $\overline{\theta} = 0$, there are strong sidelobe targets at $\overline{\theta}_1 = -0.3$, $\overline{\theta}_2 = +0.1$, $\overline{\theta}_3 = +0.25$, as shown Figure 5.6, where a normalized angle is defined as

$$\overline{\theta} \triangleq \frac{d}{\lambda}\sin\theta \qquad (5.27)$$

where d is the interelement spacing of the ULA and λ is the operating wavelength (consistent units and narrowband operation assumed).

The presence of these targets (possibly large clutter discretes) could have been previously detected, and thus their directions known. Their strong sidelobes could potentially mask weaker mainlobe targets. With this knowledge, it is desired to minimize any energy from these targets leaking into the mainbeam detection of the target of interest by nulling on transmit; that is, by placing

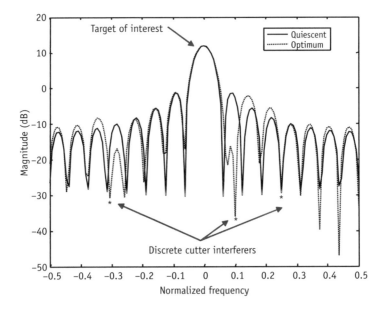

Figure 5.6 Proactive sidelobe target blanking on transmit achieved by maximizing the SCR. Note the presence of nulls in the directions of competing targets while preserving the desired mainbeam. (After: [4].)

transmit antenna pattern nulls in the directions of the unwanted targets.

For the case at hand, the (m,n)-th elements of the target and interferer transfer matrices are given, respectively, by

$$[H_T]_{m,n} = e^{j\varphi} \text{ (const.)} \tag{5.28}$$

$$[H_c]_{m,n} = \alpha_1 e^{j2\pi(m-n)\bar{\theta}_1} + \alpha_2 e^{j2\pi(m-n)\bar{\theta}_2} + \alpha_3 e^{j2\pi(m-n)\bar{\theta}_3} \tag{5.29}$$

where ϕ is an overall bulk delay (two-way propagation) that does not affect the solution to (5.25) and will thus be subsequently ignored, and $[H_c]_{m,n}$ denotes the (m,n)-th element of the clutter transfer matrix and consists of the linear superposition of the three target returns resulting from transmitting a narrowband signal from the nth transmit element and receiving it on the mth receive element of a ULA that utilizes the same array for transmit and receive [11, 15]. Note that in practice there would be a random relative phase between the signals in (5.29), which for convenience we have ignored but which can easily be accommodated by taking the expected value of the kernel $H_c'H_c$.

Solving (5.25) for the optimum eigenvector yields the transmit pattern that maximizes the SCR—which is the pattern also displayed in Figure 5.6. The competing target amplitudes were set to 40 dB relative to the desired target and 0 dB of diagonal loading was added to $H_c'H_c$ to improve numerical conditioning and allow for its inversion. Though this is somewhat arbitrary, it does provide a mechanism for controlling null depth—which in practice is limited by the amount of transmit channel mismatch [16]. Note the presence of transmit antenna pattern nulls in the directions of the competing targets as desired.

Next, we use the optimization framework to rigorously verify an intuitively obvious result regarding pulse shape and detecting a point target in uniform clutter; namely, that the best waveform for detecting a point target in distributed independent and identically distributed (i.i.d.) clutter is itself an impulse—that is, a waveform with maximal resolution—a well-known result originally proven by Manasse [17] using a different method.

Consider a unity point target arbitrarily chosen to be at the temporal origin. Its corresponding impulse response and transfer matrix are respectively given by

$$h_T[n] = \delta[n] \tag{5.30}$$

and

$$H_T = I_{N \times N} \tag{5.31}$$

where $I_{N \times N}$ denotes the $N \times N$ identity matrix. For uniformly distributed clutter, the corresponding impulse response is of the form

$$h_c[n] = \sum_{k=0}^{N-1} \tilde{\gamma}_k \delta[n-k] \tag{5.32}$$

where $\tilde{\gamma}_i$ denotes the complex reflectivity random variable of the clutter contained in the ith range cell (i.e., fast-time tap). The corresponding transfer matrix is given by

$$\tilde{H}_c = \begin{bmatrix} \tilde{\gamma}_0 & 0 & 0 & \cdots & 0 \\ \tilde{\gamma}_1 & \tilde{\gamma}_0 & & & \\ \tilde{\gamma}_2 & \tilde{\gamma}_1 & \tilde{\gamma}_0 & & \\ \vdots & & & \ddots & \\ \tilde{\gamma}_{N-1} & \tilde{\gamma}_{N-2} & \tilde{\gamma}_{N-3} & \cdots & \tilde{\gamma}_0 \end{bmatrix} \tag{5.33}$$

Assuming that the random clutter coefficients $\tilde{\gamma}_i$ are i.i.d., we have

$$E\left\{ \tilde{\gamma}_i^* \tilde{\gamma}_j \right\} = P_c \delta[i-j] \tag{5.34}$$

and thus

$$E\left\{ \left[\tilde{H}_c' \tilde{H}_c \right]_{i,j} \right\} = \begin{cases} 0, & i \neq j \\ (N+1-i)P_c, & i = j \end{cases} \tag{5.35}$$

where $[\ ]_{i,j}$ denotes the (i, j)-th element of the transfer matrix. Note that (5.35) is also diagonal (and thus invertible), but with nonequal diagonal elements.

Finally, substituting (5.31) and (5.35) into (5.26) yields

$$E\left\{\tilde{H}'_c\tilde{H}_c\right\}^{-1}\mathbf{s} = \lambda\mathbf{s} \qquad (5.36)$$

where

$$E\left\{\tilde{H}'_c\tilde{H}_c\right\}^{-1} = \frac{1}{P_c}\begin{bmatrix} d_1 & 0 & \cdots & 0 \\ 0 & d_2 & & \\ & & \ddots & \\ 0 & & \cdots & d_N \end{bmatrix} \qquad (5.37)$$

and where

$$d_i \overset{\Delta}{=} (N+i-1)^{-1} \qquad (5.38)$$

It is readily verified that the solution to (5.36) yielding the maximum eigenvalue is given by

$$\mathbf{s} = \begin{bmatrix} 1 \\ 0 \\ \vdots \\ 0 \end{bmatrix} \qquad (5.39)$$

That is, the optimum pulse shape for detecting a point target is itself an impulse. This should be immediately obvious since it is the shape that only excites the range bin with the target and zeros out all other range bin returns that contain competing clutter.

Of course transmitting a short pulse (much less an impulse) is problematic in the real world (e.g., high-peak power pulses) and thus an approximation to a short pulse in the form of a spread spectrum waveform (e.g., LFM) is often employed [10]. This example also makes clear that in the case of uniform random clutter, there is nothing to be gained by sophisticated pulse shaping for a point target other than to maximize bandwidth (i.e., range resolution). The interested reader is referred to [4] for further examples of optimizing other DoF (such as angle-Doppler) for the clutter mitigation problem, including STAP on transmit [18].

5.3 Optimum MIMO Radar for Target ID

In this section we derive the MIMO radar configuration for optimum target ID. Consider the problem of determining target type when two possibilities exist (the multitarget case is addressed later in this section). This can be cast as a classical binary hypothesis testing problem [2]; that is,

$$
\begin{array}{lll}
\text{(Target 1)} & H_1: & y_1 + n = H_{T_1} s + n \\
\text{(Target 2)} & H_2: & y_2 + n = H_{T_2} s + n
\end{array}
\tag{5.40}
$$

where H_{T_1}, H_{T_2} H denote the target transfer matrices for targets 1 and 2, respectively. For the AGN case, the well-known optimum receiver decision structure consists of a bank of matched filters, each tuned to a different target assumption and followed by comparator, as shown in Figure 5.7 [2]. Note that the above presupposes that either Target 1 or 2 is present—but not both. Also, it has been tacitly assumed that a target-present test has been conducted to ensure that a target is indeed present (i.e., binary detection test [2]). Alternatively, the null hypothesis (no target present) can be included in the test as a separate hypothesis (see discussion below for the multitarget case).

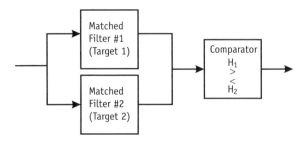

Figure 5.7 Optimal receiver structure for the binary (two target) hypothesis testing AGN problem.

Figure 5.8 illustrates the situation at hand. If Target-1 is present, the observed signal $\mathbf{y}_1 + \mathbf{n}$ will tend to cluster about the #1 point in observation space—which could include any number of dimensions relevant to the target ID problem (e.g., fast-time, angle, Doppler, polarization). The uncertainty sphere (generally ellipsoid for AGCN case) surrounding #1 in Figure 5.7 represents the 1-sigma probability for the additive noise $\mathbf{n}$—similarly for #2. Clearly, if $\mathbf{y}_1$ and $\mathbf{y}_2$ are relatively well separated, the probability of correct classification is commensurately high.

Of significant note is the fact that $\mathbf{y}_1$ and $\mathbf{y}_2$ depend on the transmit signal $\mathbf{s}$, as shown in (5.40). Consequently, it should be possible to select an $\mathbf{s}$ that maximizes the separation between $\mathbf{y}_1$ and $\mathbf{y}_2$, thereby maximizing the probability of correct classification under modest assumptions regarding the conditional probability density functions (pdfs) (e.g., unimodality, see below); that is,

$$\max_{\{\mathbf{s}\}} \; |\mathbf{d}'\mathbf{d}| \qquad (5.41)$$

where

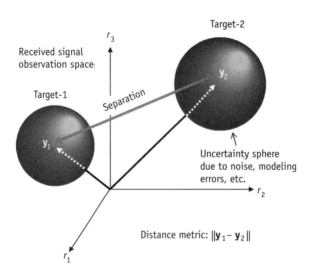

Figure 5.8   The two-target ID problem.

$$\mathbf{d} \stackrel{\Delta}{=} \mathbf{y}_1 - \mathbf{y}_2$$
$$= H_{T_1}\mathbf{s} - H_{T_2}\mathbf{s}$$
$$= \left(H_{T_1} - H_{T_2}\right)\mathbf{s} \qquad (5.42)$$
$$\stackrel{\Delta}{=} H\mathbf{s}$$

where

$$H \stackrel{\Delta}{=} H_{T_1} - H_{T_2} \qquad (5.43)$$

Substituting (5.42) into (5.41) yields

$$\max_{\{\mathbf{s}\}} \quad |\mathbf{s}'H'H\mathbf{s}| \qquad (5.44)$$

which is precisely of the form (5.10), and thus has a solution yielding maximum separation given by

$$(H'H)\mathbf{s}_{opt} = \lambda_{max}\mathbf{s}_{opt} \qquad (5.45)$$

Equation (5.45) has an interesting interpretation: $\mathbf{s}_{opt}$ is the transmit input that maximally separates the target responses and is the maximum eigenfunction of the transfer kernel $H'H$ formed by the difference between the target transfer matrices (i.e., [5.43]). Again, if the composite target transfer matrix is stochastic, $H'H$ is replaced with its expected value $E\{H'H\}$ in (5.45). Next, we illustrate with a two target numerical example.

Let $h_1[n]$ and $h_2[n]$ denote the impulse responses of targets #1 and #2, respectively, as shown in Figure 5.9. Figure 5.10 shows two different (normalized) transmit waveforms: (1) chirp, and (2) optimum (per [5.45])—along with their corresponding normalized separation norms of 0.45 and 1, respectively (which corresponds to 6.9 dB improvement in separation). To determine the relative probabilities of correct classification for the different transmit waveforms, one would first need to set the SNR level (which fixes the conditional pdfs herein assumed to be circular Gaussian), then measure the amount of overlap to calculate the probability [2].

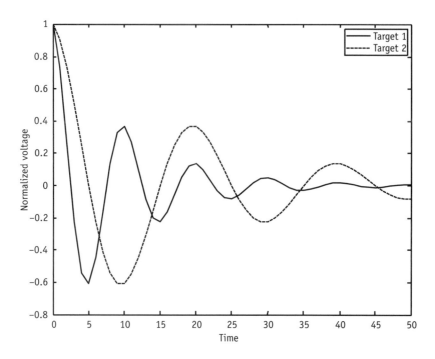

Figure 5.9 Target impulse responses (Green's functions) utilized for the two-target identification problem.

It is not at all obvious why the optimum pulse achieves much better performance simply by looking at the time domain waveform. However, an examination of Figure 5.11 reveals the mechanism by which enhanced separation is achieved. It shows the Fourier spectrum of $H(\omega) = H_{T_1}(\omega) - H_{T_2}(\omega)$ along with that of $S_{opt}(\omega)$. Note that $S_{opt}(\omega)$ places more energy in those spectral regions where $H(\omega)$ is large (i.e., regions where the difference between targets is large—again an intuitively appealing result). While pulse modulation was used to illustrate the optimum transmit design equations, we could theoretically have used any transmit DoF (e.g., polarization). The choice clearly depends on the application at hand.

Referring to Figure 5.8, we see that the solution to (5.45) maximizes the conditional pdf separation metric $d = \|\mathbf{d}\|$. For the additive noise case with unimodal pdf, maximizing d minimizes the overlap of the two conditional pdfs, and thus is a statistically optimum choice for $\mathbf{s}$. Note that a Gaussian assumption is not required.

The above can be readily extended to the multitarget case. Given L targets in general, we wish to ensure that the L-target response

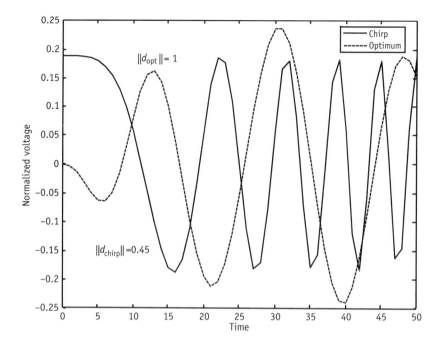

Figure 5.10 LFM and optimum pulse waveforms.

spheres are maximally separated (a type of inverse sphere packing problem [19]). To accomplish this, we would like to jointly maximize the norms of the set of separations $\{\|\mathbf{d}_{ij}\| \mid i = 1: L; j = i + 1: L\}$; that is,

$$\max_{\mathbf{s}} \sum_{i=1}^{L} \sum_{j=i+1}^{L} \left| \mathbf{d}'_{ij} \mathbf{d}_{ij} \right| \tag{5.46}$$

Since, by definition, $\mathbf{d}_{ij}$ is given by

$$\mathbf{d}_{ij} \overset{\Delta}{=} \left(H_{T_i} - H_{T_j} \right) \mathbf{s}$$

$$\overset{\Delta}{=} H_{ij} \mathbf{s} \tag{5.47}$$

(5.46) can be rewritten as

$$\max_{\mathbf{s}} \; \mathbf{s}' \left(\sum_{i=1}^{L} \sum_{j=i+1}^{L} H'_{ij} H_{ij} \right) \mathbf{s} \overset{\Delta}{=} \mathbf{s}' K \mathbf{s} \tag{5.48}$$

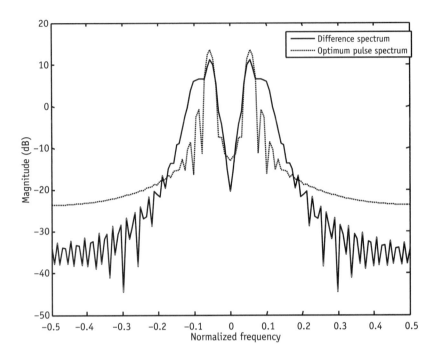

Figure 5.11 Comparison of optimum pulse and target difference spectra illustrating the mechanism by which enhanced target separation is achieved.

Since $K \in \mathbb{C}^{N \times N}$ is the sum of positive semidefinite matrices, it shares this same property and thus the optimum transmit input satisfies

$$Ks_{opt} = \lambda_{max} s_{opt} \qquad (5.49)$$

Figure 5.12 depicts the impulse responses of three different targets, two of which are the same as in the previous two target example. Solving (5.48)–(5.49) yields an optimally separating waveform whose average separation defined by (5.46) is 1.0, as compared to 0.47 for the chirp—an improvement of 6.5 dB, which is slightly less than the previous example as would be expected since more is being asked of the waveform probe for the three target case. As expected, the optimum waveform outperforms (in this case significantly) that of an unoptimized pulse such as the chirp.

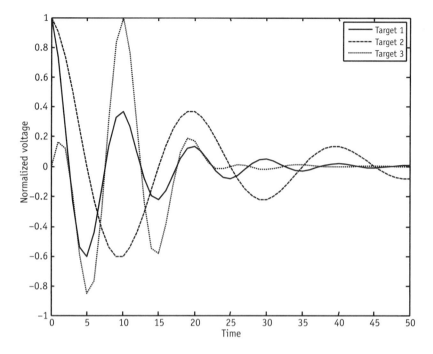

Figure 5.12 Target impulse responses utilized for the three-target identification problem.

References

[1] Greenberg, M. D., *Applications of Green's Functions in Science and Engineering,* New York: Prentice Hall, 1971.

[2] Van Trees, H. L., *Detection, Estimation and Modulation Theory. Part I,* New York: Wiley, 1968.

[3] Barton, D. K., *Modern Radar System Analysis,* Norwood, MA: Artech House, 1988.

[4] Guerci, J. R., *Cognitive Radar: The Knowledge-Aided Fully Adaptive Approach,* Norwood, MA: Artech House, 2010.

[5] Papoulis, A., *Signal Analysis,* New York: McGraw-Hill, 1984.

[6] Papoulis, A., *Circuits and Systems: A Modern Approach,* New York: Holt, Rinehart and Winston, 1980.

[7] Horn R. A., and C. R. Johnson, *Matrix Analysis.* Cambridge, UK: Cambridge University Press, 1990.

[8] Pierre, D. A., *Optimization Theory with Applications*: Mineola, NY: Dover Publications, 1986.

[9]   Guerci, J. R., and S. U. Pillai, "Theory and Application of Optimum Trans-mit-Receive Radar," in *The Record of the IEEE 2000 International Radar Conference*, pp. 705–710.

[10] . Cook, C. E, and M. Bernfeld, *Radar Signals*, New York: Academic Press, 1967.

[11] Richards, M. A., *Fundamentals of Radar Signal Processing*, New York: McGraw-Hill, 2005.

[12] Lindenfeld, M. J., "Sparse Frequency Transmit-and-Receive Waveform Design," *IEEE Transactions on Aerospace and Electronic Systems*, Vol. 40, 2004, pp. 851–861.

[13] Van Trees, H. L., *Detection, Estimation, and Modulation Theory*, Part II, New York: John Wiley and Sons, 1971.

[14] Van Trees, H. L., *Detection, Estimation, and Modulation Theory: Radar-Sonar Signal Processing and Gaussian Signals in Noise, Part III*, Krieger Publishing Co., Inc., 1992.

[15] Guerci, J. R., *Space-Time Adaptive Processing for Radar*, Norwood, MA: Artech House, 2003.

[16] Monzingo, R. A., and T. W. Miller, *Introduction to Adaptive Arrays*, Raleigh, NC: SciTech Publishing, 2003.

[17] Manasse, R., "The Use of Pulse Coding to Discriminate Against Clutter," *Defense Technical Information Center (DTIC)*, Vol. AD0260230, June 7, 1961.

[18] Guerci, J. R., *Space-Time Adaptive Processing for Radar*, Second Edition, Norwood, MA: Artech House, 2014.

[19] Hsiang, W. Y., "On the Sphere Packing Problem and the Proof of Kepler's Conjecture," *International Journal of Mathematics*, Vol. 4, 1993, pp. 739–831.

6

Adaptive MIMO Radar and MIMO Channel Estimation

In this chapter, we introduce the concept of adaptive MIMO radar. In Chapter 5 we introduced optimum MIMO radar, which required knowledge of the relevant radar channel (targets, clutter, interference). In real-world applications, a priori knowledge of the channel is only approximate at best or not known at all. Thus, techniques for approximating optimal techniques using adaptive methods are required.

In Section 6.1, we introduce adaptive MIMO techniques that are essentially extensions of methods employed in conventional adaptive radar such as STAP, which are also attempts to approximate optimal receiver techniques [1–3]. In Section 6.2 we introduce a new approach to channel estimation that combines orthogonal MIMO techniques with optimum MIMO to create an entirely new hybrid approach to radar operation.

6.1 Introduction to Adaptive MIMO Radar

Chapter 5 derived the optimal multidimensional transmit-receive (i.e., MIMO) design equations that assumed exact knowledge (deterministic and/or statistical) of the channel (target and interference). However, as those familiar with real-world radar are well aware, channel characterization in large part must be performed on-the-fly; that is to say, adaptively. This is simply a result of the dynamic nature of real-world targets and especially interference.

While a plethora of techniques have been developed for radar receiver adaptivity, estimating requisite channel characteristics for adapting the transmit function—especially for transmit-dependent interference such as clutter—is a relatively new endeavor. In Section 6.2, we explore several approaches for addressing the adaptive MIMO optimization problem.

A classic example of transmit signal-independent interference is additive noise jamming [4]. For the case where no a priori knowledge is available, the baseline method of sample covariance estimation (and its many variants such as diagonal loading and principal components [2, 5, 6]) is often utilized. In addition to its statistical optimality properties (it is the maximum likelihood solution for the i.i.d. additive Gaussian noise case [7]), efficient parallel processing implementations have been developed facilitating its real-time operation [8].

Figure 6.1 depicts a common procedure for estimating additive, transmit-independent interference statistics. Specifically, the interference covariance matrix $R \in \mathbb{C}^{N \times N}$ is approximated by $\hat{R} \in \mathbb{C}^{N \times N}$, where

$$\hat{R} = \frac{1}{L} \sum_{q \in \Omega} \mathbf{x}_q \mathbf{x}_q' \tag{6.1}$$

where $\mathbf{x}_q \in \mathbb{C}^N$ denotes the N-dimensional receive array snapshot (spatial, spatiotemporal, etc.) corresponding to the qth independent temporal sample (e.g., a range or Doppler bin), and L denotes the number of i.i.d. samples selected from a suitable set of training samples Ω to form the summation. As depicted in Figure 6.1, this training region is often chosen to be close in range to the range cell of interest (though there are many variants of this). If, moreover, the selected samples are Gaussian and i.i.d., then (6.1) can be shown to provide the maximum likelihood estimate of [7]. We illustrate this approach in the following example.

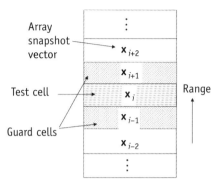

Figure 6.1 Illustration of a common method for estimating the interference statistics for the additive transmit-independent case. (After: [9].)

This is a repeat of the previous multipath interference example from Chapter 5 with the notable exception of unknown interference statistics that must be estimated on the fly. As a consequence, an estimate of the covariance matrix is used for the whitening filter rather than the actual covariance, as was the case in Chapter 5.

Plotted in Figure 6.2 is the overall output SINR loss relative to the optimum for the short-pulse case as a function of the number of independent samples used in (6.1). The results shown were based

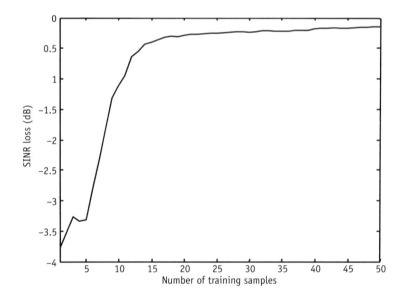

Figure 6.2 Effect of sample support on output SINR loss for the multipath interference scenario. (After: [9].)

on 50 Monte Carlo trials (rms average) with a JNR of 50 dB and a small amount of diagonal loading to allow for inversion when the number of samples is less than 11 (positive semidefinite case).

It is interesting to note the rapid convergence and contrast this with SINR loss performance for adaptive beamforming [10]—which is generally significantly slower. This is due to the fact that we are only estimating the single dominant eigenvalue/eigenvector pair. For an authoritative examination of principal components estimation and convergence properties, the interested reader is referred to [11].

6.2 MIMO Channel Estimation Techniques

In [9] it was first realized that orthogonal MIMO techniques have another very useful role in support of optimum MIMO; namely, they can be used to estimate the channel in a rapid and efficient manner. Since then, a number of techniques have been developed that exploit this capability (see for example [12–14]). However, one of the earliest real-world experiments of MIMO channel estimation was performed by Coutts et al. [15].

The experiment described in [15] is illustrated in Figure 6.3. It depicts an airborne high-value target (HVT) that can be detected

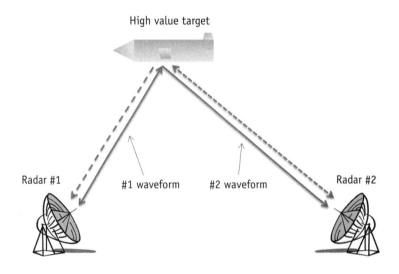

Figure 6.3   MIMO cohere-on-target approach for maximizing distributed radar performance [15]. (After: [9].)

simultaneously by two geographically disparate radars. Given the HVT nature of the target, it is desired to have the two radars work coherently in order to maximize the overall SNR at each radar. To achieve on-target coherency, the two waveforms from each radar need to interfere constructively. However, to accomplish this requires precise knowledge of the transmit pathways to a fraction of a wavelength [15]—essentially a dynamic *channel* calibration, where the channels consist of the propagation and target response for the two radars.

The requisite relative time delays between the two radars (as seen by the target) can be estimated by simultaneously transmitting orthogonal waveforms, which are then detected and processed in each radar as follows:

- At each radar, the known one-way time delay to the target is subtracted from the total transit time for the sister radar (precise time synchronization is assumed). The remaining time delay is thus due to the first leg of the bistatic path (see Figure 6.3).

- By precompensating a joint waveform in each radar, the two waveforms can be made to cohere on the target, resulting in a 3-dB SNR boost (ideally). If the above procedure is repeated for N radars, as much as a $10\log N$, decibel gain in SNR is theoretically achievable.

While relatively straightforward to describe, the above procedure is replete with many real-world difficulties, including target motion compensation to a fraction of a wavelength and precise phase/timing stability. The reader is referred to [15] for further details.

As mentioned previously, the orthogonal waveform MIMO radar approach can provide a means for adaptively estimating the composite target-interference channel since the individual input-output responses can, under certain circumstances, be resolved simultaneously. However, once an estimate of the composite channel is achieved, the optimal MIMO transmit-receive configuration derived in Chapter 5 should be employed to maximize SINR, SCNR, or probability of correct classification.

To see how these advanced MIMO probing techniques work, consider a spatial clutter transfer function H for a given range bin of interest; that is,

$$H = \begin{bmatrix} H_{11} & H_{12} & \cdots & H_{1N} \\ H_{21} & H_{22} & & \\ \vdots & & \ddots & \\ H_{N1} & & & H_{NN} \end{bmatrix} \qquad (6.2)$$

where, without loss of generality, we have assumed the same number of transmit and receive channels. When we use a conventional phased array approach (i.e., single-input, multiple output [SIMO]), we cannot probe the individual transmit-receive pathways for each transmitter element—and thus cannot fully characterize the clutter channel. However, using a MIMO approach it is possible to do just that.

The associated transmit signal vector $\mathbf{s}$ would be of the form

$$\mathbf{s} = \begin{bmatrix} \mathbf{s}_1 \\ \vdots \\ \mathbf{s}_N \end{bmatrix} \qquad (6.3)$$

where $\mathbf{s}_i$, for example, could represent the waveform transmitted out of the ith transmitter (see [16–19] for further background on this approach). Thus, H_{ij} represents the transfer function between jth transmit element (or subarray) and the ith receive element in this example. For a basic narrowband stationary clutter model, a reasonable approximation is $H_{ij} = h_{ij}I$; that is, the clutter signal does not vary during a pulse—particularly a short one. Pulse-to-pulse variation due to relative Doppler will of course be allowed as is normally the case for MTI STAP.

Using a receiver structure similar to that utilized by CDMA [17], it is possible to reconstruct estimates of the various elements of (6.2). To see how this is possible, consider the output of the ith receive channel matched filter tuned to the jth transmit waveform; namely

$$\begin{aligned}
y_{ij} &= \mathbf{s}'_j \mathbf{z}_i \\
&= \alpha_{ij} \mathbf{s}'_j H_{ij} \mathbf{s}_j + n_{r_i} + n_i \\
&= \alpha_{ij} \mathbf{s}'_j \left(h_{ij} I \right) \mathbf{s}_j + n_{r_i} + n_i \\
&= \alpha_{ij} h_{ij} + n_{r_i} + n_i
\end{aligned} \tag{6.4}$$

where $\mathbf{z}_i$ is the received signal at the ith receiver input, n_j is the receiver noise in the jth receiver channel, α_{ij} is a scalar free-space propagation factor that subsumes the transmit power and R^4 losses, and n_{r_j} is the residue from the CDMA processing. Note that if TDMA or FDMA MIMO coding was used, this latter noise term would be zero. However, these other techniques have their pros and cons, such as sampling the channel at different times (TDMA), and frequencies (FDMA)—which may or may not be acceptable depending on the application. Note also that we have assumed a perfectly normalized matched filter (i.e., $\mathbf{s}'_i \mathbf{s}_i = 1$). Any deviation from this would be subsumed into α_{ij}. We thus see that to within a common free-space scale factor, the output of the ith receiver is a scaled noisy estimate $\hat{h}_{ij}$ of the channel elements h_{ij}.

Recall from basic STAP theory that an estimate $\hat{R}_{sm}$ of the actual interference covariance matrix R is generally obtained (explicitly or implicitly) from a maximum likelihood (ML) optimization that leads to sample based covariance of the form

$$\hat{R}_{sm} = \kappa \sum_{k=1}^{K} \mathbf{x}_k \mathbf{x}'_k \tag{6.5}$$

where $\mathbf{x}_k$ denotes the space-time receive array snapshot vector for the kth range bin, and K denotes the total number of samples used. A major and fundamental drawback this approach is the assumption of statistical stationarity and mutual sample independence among range bins [2]. Indeed the estimate from a single range bin is identically rank-one since it is formed by a single vector outer product.

To relate (6.5) to (6.4), note that $\mathbf{x}_k$ has the form

$$\mathbf{x}_k = H_k \mathbf{s} + \mathbf{n}_k \tag{6.6}$$

where H_k is the full dimension transfer function, $\mathbf{s}$ is the transmit steering vector, $\mathbf{x}_k$ is the received signal, and $\mathbf{n}_k$ receiver noise (assumed to be the usual stationary white noise both spatially and temporally). Note that this traditional SIMO approach only yields a rank one measurement of the channel transfer function for a given range bin. This is why averaging over multiple range bins is required to obtain a full rank and ultimately useful estimate of the covariance. From (6.6) we also infer the form of the clutter component R_c of the total interference covariance matrix, $R = R_c + \sigma^2 I$; namely, $R_{c_k} = \mathrm{cov}(H_k \mathbf{s})$, for the kth range bin.

Note that knowledge of (6.6) is tantamount to knowledge of R_c for the stationary case of white noise illumination since

$$\begin{aligned} R_{c_k} &= \mathrm{cov}(H_k \mathbf{s}) \\ &= \mathrm{cov}(H_k \mathbf{n}) \\ &= E(H_k H_k') \end{aligned} \qquad (6.7)$$

where $\mathbf{n}$ is unity variance white noise and where for the static clutter case, $E(H_k H'_k) = H_k H'_k$. Of more practical significance is the fact that knowledge of H_k can be used to predict and thus coherently subtract out the clutter interference for the kth range bin. Of course to do this perfectly would require exact knowledge of both H_k and $\delta \mathbf{s}$, neither of which is possible. However, in highly inhomogeneous *and* strong clutter environments, this prefiltering stage can be used to detrend the otherwise highly nonstationary clutter series with range index [9, 20]. The remaining clutter residue would thus be more statistically stationary and in turn be more amenable to traditional sample covariance based approaches.

Let $\delta \mathbf{s}$ and δH denote the estimation errors in the transmit steering vector $\mathbf{s}$ and the clutter transfer function H, respectively. The corresponding prediction with errors would have the form

$$\begin{aligned} \hat{\mathbf{y}}_i &= (H + \delta H)(\mathbf{s} + \delta \mathbf{s}) \\ &= \mathbf{y}_i + \delta H \mathbf{s} + H \delta \mathbf{s} + \delta H \delta \mathbf{s} \\ &= \mathbf{y}_i + \delta \mathbf{y}_i \end{aligned} \qquad (6.8)$$

Making the reasonable assumption that $\delta\mathbf{s}$ and δH are zero mean and uncorrelated, the variance in the error $\delta\mathbf{y}_i$ is given by

$$\text{var}(\delta\mathbf{y}_i) = \text{var}(\delta H)\mathbf{s} + H\,\text{var}(\delta\mathbf{s}) \tag{6.9}$$

Again, this type of prefiltering (or more precisely detrending) would only be indicated in *both* strong and highly nonstationary clutter environments but could result in a far more stationary clutter residue that would be more amenable to traditional sample covariance estimation techniques.

6.3 Application to Large Clutter Discrete Mitigation

Another important application for which MIMO channel probing techniques can be applied is in the detection, estimation, and proactive (transmit-based) mitigation of large clutter discretes [13]. Large clutter discretes can effectively act as jammers inasmuch as their presence can desensitize detectors (e.g., CFAR [21]) and their sidelobes can mask weaker targets. There is thus a strong desire to detect and mitigate their impacts. The fact that the clutter discrete response is a function of the space-time transmit function means that MIMO and optimum MIMO techniques could be used to proactively mitigate, in contrast to receiver-only techniques such as STAP.

The concept of MIMO radar probing for enhanced clutter estimation was first introduced in [14]. It was shown that by probing the environment using a receiver structure similar to that utilized by CDMA, it is possible to reconstruct estimates of the various elements of composite transmit/receive channel matrix that characterizes the propagation of the radar waveform from each transmitter antenna array element and each receiver element. This in turn resulted in an overall reduction in the number of training samples required for adequate STAP performance—a useful result in nonstationary clutter environments. In this section we will show how this probing concept can be extended in a GMTI radar system to more rapidly extract information about strong clutter discretes in the radar scene, and then proactively mitigate them by the use of an appropriate space-time waveform that notches the illumination pattern in the direction of the strongest discretes.

A key feature of the new MIMO clutter probing waveforms is the exploitation of the strong signal characteristics (by definition) of clutter discretes. This allows for the use of extremely short pulses, which in turn helps to ameliorate the cross-correlation integrated range sidelobes clutter leakage discussed in [18] that is inevitable using longer pulse compression waveforms in distributed GMTI clutter environments (see Figure 6.4). The short pulses also allow for shorter PRIs, which in turn can be used to minimize overhead impacts to the radar scheduler.

Consider an N-channel ULA with N coincident transmit and receive antenna elements. Furthermore, the radar is capable of transmitting MIMO waveforms in some manner. This can be accomplished for example via the use of separate waveform generators for each transmit phase center (costly) or by a suitable phase modulation such as a biphase modulator in each transmit channel (less costly).

Let $\{s_1, ..., s_N\}$ denote the N short-pulse MIMO probing waveforms. Thus, the total composite space-time transmit waveform $\mathbf{s}$ is of the form

$$\mathbf{s} = \begin{bmatrix} \mathbf{s}_1 \\ \vdots \\ \mathbf{s}_N \end{bmatrix} \tag{6.10}$$

The exact dimension of $\mathbf{s}$ depends on the choice of transmit DoF. For example, $\mathbf{s}_N$ could be the fast-time single polarization

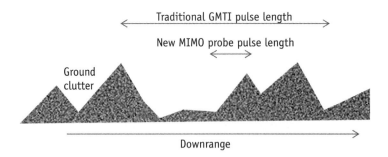

Figure 6.4 New MIMO probing waveform can use much shorter pulses than traditional GMTI pulse compression waveforms, which ameliorates cross-correlation range sidelobe clutter leakage.

waveform transmitted out of the nth transmit subarray or antenna. Unlike the channel probing approach introduced in [14], we will not attempt to estimate the clutter transfer function, but rather will use CDMA-like matched filtering in the receiver to simultaneously form transmit-receive beams that cover the desired radar field-of-view (FoV) of interest. The maximum extent of this FoV is determined by the width of the spoiled MIMO transmit beamwidth that in turn is dependent on the individual element or subarray patterns.

While it is possible to synthesize the final output MIMO transmit-receive beams in one batch processing step using very large matched filters, we will use a sequential process that is easier to describe and in some cases implement. Let $\{\mathbf{y}_1, \ldots, \mathbf{y}_N\}$ denote the set of fast-time waveforms received in each receive element or subarray. For the mth range bin, and ith receive channel, N transmit DoF can be reconstructed using CDMA style matched filtering. That is, applying the matched filter weight vector $\mathbf{w}_j$, given by

$$\mathbf{w}_j = \mathbf{s}_j \qquad (6.11)$$

to the ith receive channel signal $\mathbf{y}_i$ yields the complex scalar residue z_{ij} given by

$$z_{ij} = \mathbf{w}_j' \mathbf{y}_i \qquad (6.12)$$

If $\mathbf{a}_\theta$ denotes an N-dimensional transmit steering vector of interest, then the N-dimensional received signal at the the mth range bin, and ith receive channel, can be focused by forming

$$x_i = \mathbf{a}_\theta' \mathbf{z}_i \qquad (6.13)$$

where x_i is the complex scalar output of the ith receive channel for the mth range bin, and $\mathbf{z}_i$ denotes the N-dimensional vector whose elements are given by (6.12). Lastly, a total focused transmit receive beam is achieved by applying an N-dimensional receive steering vector $\mathbf{b}_\theta$; that is,

$$r = \mathbf{b}_\theta' \mathbf{x} \qquad (6.14)$$

where x_i denotes the N-dimensional vector whose elements are given by (6.13). If the receive and transmit array antenna manifolds and electrical channels are equal, $b_\theta = a_\theta$.

Since all of the above functions are performed digitally in the receiver, it is possible to simultaneously synthesize a set of focused MIMO transmit-receive beams that covers the entire FoV in a highly parallel processing implementation. The resulting range-angle clutter maps can then be quickly scanned (again in parallel) for large clutter discretes. Though there is a loss in CNR due to the use of short pulses and a spoiled MIMO transmit pattern, this does not present a problem since the goal is the detection of very large clutter discretes. Indeed, in a very real sense this intentional desensitization ensures that only truly strong discretes will be detected in this fashion.

Again, our goal is to achieve large clutter discrete detection with as minimum an impact to radar timeline and resources as possible. The use of very short transmit pulses and MIMO probing allows for the simultaneous and rapid probing of the entire FoV using a small fraction of a traditional GMTI coherent processing interval (CPI) [1].

Lastly, once large clutter discretes are detected and localized in range-angle space, any number of knowledge-aided (KA) techniques can be employed to mitigate or reduce their deleterious impact. Such KA techniques vary in complexity from simple intelligent blanking to more sophisticated KA nulling on both receive and possibly transmit [14].

Consider a standoff airborne radar with a 10-element antenna array with uniform half-wavelength element spacing. The radar is capable of transmitting a unique waveform on all 10 antenna elements. The radar operating frequency is 1,240 MHz and the bandwidth is 20 MHz. In the examples shown below, the radar transmits random phase coded waveforms on each antenna element (i.e., $s_n(t) = \exp(j\phi_n(t))$) where $\phi_n(t)$ is uniformly distributed on $[0,2\pi]$). Further radar parameters are contained in Table 6.1.

We begin with an example involving three strong clutter discretes. The clutter probing response for the MIMO processing approach given in (6.5) is shown in Figure 6.5. We see that three strong clutter discretes are clearly resolved and visible in the output. As a comparison, Figure 6.6 shows the output when only one channel is

Table 6.1
Radar Parameters

Parameter	Value
Transmitter peak power	1 kW
Duty factor	0.1
Transmit ant. gain	18 dB
Receive ant. gain	18 dB
Target RCS	10 dB sm
Wavelength	0.24m
Radar system losses	5 dB
Noise temp.	290K
PRF	1000 Hz
Bandwidth	10 MHz
Noise factor	5 dB

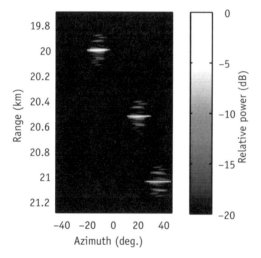

Figure 6.5 MIMO probing response for three large clutter discrete. (©2015 IEEE. Reprinted, with permission, from *Proceedings of the 2015 IEEE Radar Conference.*)

used to transmit. This represents a more traditional beam-spoiling approach to illuminating a wide area [22].

We see that, as expected, the azimuth response of the single waveform case has much higher sidelobes since, unlike the MIMO case, it does not have sufficient DoF to enable the formation of a two-way antenna pattern in the signal processor. Figure 6.7 provides an azimuth cut for the clutter discrete at 20-km range. This result clearly shows the lower sidelobe response of the SPP MIMO

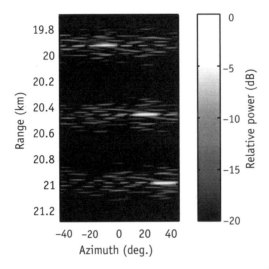

Figure 6.6 Conventional SIMO (non-MIMO) results. (©2015 IEEE. Reprinted, with permission, from *Proceedings of the 2015 IEEE Radar Conference.*)

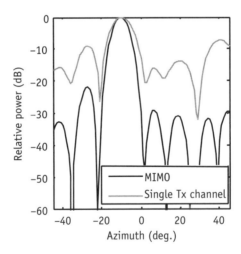

Figure 6.7 Azimuth response for clutter discrete. (©2015 IEEE. Reprinted, with permission, from *Proceedings of the 2015 IEEE Radar Conference.*)

probing approach. It also shows that the MIMO approach results in a narrower mainbeam response, which is expected given that the MIMO approach allows for the formation of a two-way pattern. The lower sidelobes and narrower mainbeam response will be important for resolving clutter in scenarios involving a high density of strong clutter discretes such as scenarios characterized by highly heterogeneous terrain.

We anticipate that the MIMO probing waveforms will be inter-leaved/embedded with the actual radar data. As an example, a short MIMO pulse could be added to the end of each radar pulse in a coherent processing interval. Once the clutter discretes are detected using the MIMO probing pulses, the collected MIMO data itself can be used to estimate a spatial filter to null the clutter discretes when processing the main radar pulses. The following analysis shows that it is feasible to use the MIMO probe data as secondary training data to support adaptive spatial nulling of clut-ter discretes in the scene.

The radar model shown in Table 6.1 was used to compute the discrete-to-noise ratio (DNR) as a function of range and azimuth angle for a clutter discrete with a 40-dB radar cross section. The result is shown in Figure 6.8 where we see that, as expected, strong sidelobe discretes can result in very strong and thus easily detected radar returns that will lead to high false alarm rates.

Figure 6.9 shows a similar calculation for the MIMO probing pulses. In this case we show the DNR on a single MIMO transmit/ receive channel after pulse compression and Doppler processing. In this case the transmit and receive antenna patterns are omnidi-rectional in the horizontal dimension with gains that are one-tenth of the full antenna gain. Also, we have assumed that the MIMO waveforms on each pulse have one-tenth the pulse width of the main radar pulses resulting waveform duty factor that is 10 dB

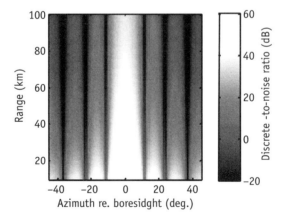

Figure 6.8   Discrete-to-noise ratio for a clutter discrete with a 40-dB radar cross section. (©2015 IEEE. Reprinted, with permission, from *Proceedings of the 2015 IEEE Radar Conference.*)

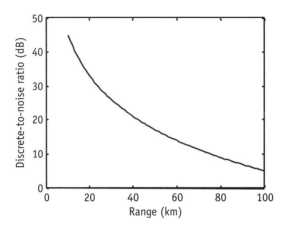

Figure 6.9   Discrete-to-noise ratio for the MIMO probing mode. Result is for a single MIMO transmit/receive channel. (©2015 IEEE. Reprinted, with permission, from *Proceedings of the 2015 IEEE Radar Conference.*)

lower than the main radar mode. We see that the clutter discrete (40-dB RCS in this case), as a result of its high RCS, still results in a strong received signal despite the loss in gain and transmit. We note that a final step in the processing would be performed to produce the beamformed output prior to detection of the clutter discretes in the scene.

Once the discretes are detected the MIMO data provides sufficient samples to produce a full rank estimate of the spatial covariance of the clutter discrete. This is possible because, assuming low cross correlation between the MIMO waveforms, each transmit channel produces an independent receive spatial snapshot for the clutter discrete. In this example, this would result in 10 spatial samples that can be used to estimate the 10×10 spatial covariance of the discrete. This covariance estimate can be used in the usual manner [23] to compute an adaptive spatial filter that is then applied to the main radar data at the range/Doppler bin of the detected clutter discrete. It is important to note that this processing results in a nulling of the clutter discrete as opposed to blanking [24], which is a much more desirable result since it has the potential to maintain a high probability of target detection in the range-Doppler bin of interest.

Figure 6.10 shows the DNR after spatial filtering using the MIMO probe data as discussed above. By comparing this result with Figure 6.8, we see that the MIMO data provides sufficient

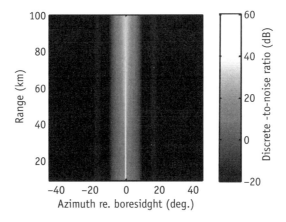

Figure 6.10 Discrete-to-noise ratio after adaptive spatial filtering using the MIMO probe data. (©2015 IEEE. Reprinted, with permission, from *Proceedings of the 2015 IEEE Radar Conference.*)

data (numbers of samples and signal-to-noise ratio) to null the sidelobe discretes to below the noise floor over most of the range extent. The approach also provides some nulling of the discretes in the skirts of the mainbeam. This is a very encouraging result in that it shows that significant performance gains are potentially possible using a probe, learn, adapt methodology. We note that this result likely represents an upper bound on performance given that we have ignored the effects of finite cross correlation between the MIMO waveforms. Also, we have ignored the impact of clutter. In a final implementation, the clutter discrete spatial filter would need to be appropriately incorporated into the particular STAP algorithm being used on the system. Incorporating and characterizing the impact of these effects is an area for future work.

6.4 Summary

In this chapter, we showed that MIMO techniques can play a new role in both channel estimation and as an integral part of adaptive *optimum* MIMO. The optimum MIMO techniques introduced in Chapter 5 assumed knowledge (deterministic or statistical) of the relevant channel components (target[s], clutter, jamming), an assumption often not justified in practice. However, the inherent diversity probing nature of orthogonal MIMO introduced in Chapters 3 and 4 can provide the requisite channel information adaptively. In this chapter, we also illustrated the MIMO channel probing technique applied to an important strong clutter discrete mitigation problem that often arises in many radar applications, especially GMTI.

References

[1] Guerci, J. R., *Space-Time Adaptive Processing for Radar,* Second Edition, Norwood, MA: Artech House, 2014.

[2] Guerci, J. R., *Space-Time Adaptive Processing for Radar*, Norwood, MA: Artech House, 2003.

[3] Ward, J., "Space-Time Adaptive Processing for Airborne Radar," *Space-Time Adaptive Processing (Ref. No. 1998/241), IEE Colloquium on*, p. 2, 1998.

[4] Monzingo, R. A., and T. W. Miller, *Introduction to Adaptive Arrays*, Raleigh, NC: SciTech Publishing, 2003.

[5] Carlson, B. D., "Covariance Matrix Estimation Errors and Diagonal Loading in Adaptive Arrays," *Aerospace and Electronic Systems, IEEE Transactions on*, Vol. 24, 1988, pp. 397–401.

[6] Haimovich, A. M., and M. Berin, "Eigenanalysis-Based Space-Time Adaptive Radar: Performance Analysis," *Aerospace and Electronic Systems, IEEE Transactions on*, Vol. 33, 1997, pp. 1170–1179.

[7] Van Trees, H. L., *Detection, Estimation and Modulation Theory.* Part I. New York: Wiley, 1968.

[8] Farina, A., and L. Timmoneri, "Real-Time STAP Techniques," *Electronics & Communication Engineering Journal*, Vol. 11, 1999, pp. 13–22.

[9] Guerci, J. R., *Cognitive Radar: The Knowledge-Aided Fully Adaptive Approach.* Norwood, MA: Artech House, 2010.

[10] Reed, I. S., J. D. Mallett, and L. E. Brennan, "Rapid Convergence Rate in Adaptive Arrays," *Aerospace and Electronic Systems, IEEE Transactions on*, Vol. AES-10, 1974, pp. 853–863.

[11] Smith, S. T., "Covariance, Subspace, and Intrinsic Cramer-Rao Bounds," *Signal Processing, IEEE Transactions on*, Vol. 53, 2005, pp. 1610–1630.

[12] Guerci, R., J. S. Bergin, M. Khanin, and M. Rangaswamy, "A New MIMO Clutter Model for Cognitive Radar," in *2016 IEEE Radar Conference (RadarConf)*, pp. 1–6.

[13] Bergin, J. S., J. R. Guerci, R. M. Guerci, and M. Rangaswamy, "MIMO Clutter Discrete Probing For Cognitive Radar," presented at the *IEEE International Radar Conference*, Arlington, VA, 2015.

[14] Guerci, J. R., R. M. Guerci, M. Ranagaswamy, J. S. Bergin, and M. C. Wicks, "CoFAR: Cognitive Fully Adaptive Radar," presented at the *IEEE Radar Conference*, Cincinnati, OH, 2014.

[15] Coutts, S., K. Cuomo, J. McHarg, F. Robey, and D. Weikle, "Distributed Coherent Aperture Measurements for Next Generation BMD Radar," in *Fourth IEEE Workshop on Sensor Array and Multichannel Processing*, 2006, pp. 390–393.

[16] Bliss, D. W., "Coherent MIMO Radar," presented at the *International Waveform Diversity and Design Conference (WDD)*, 2010.

[17] Bliss, D. W., and K. W. Forsythe, "Multiple-Input Multiple-Output (MIMO) Radar and Imaging: Degrees of Freedom and Resolution," presented at the Conference Record of the Thirty-Seventh Asilomar Conference on Signals, Systems and Computers, 2003.

[18] Bliss, D. W., K. W. Forsythe, S. K. Davis, et al., "GMTI MIMO Radar," presented at the *International Waveform Diversity and Design Conference*, 2009.

[19] Forsythe, K. W., D. W. Bliss, and G. S. Fawcett, "Multiple-Input Multiple-Output (MIMO) Radar: Performance Issues," in *Conference Record of the Thirty-Eighth Asilomar Conference on Signals, Systems, and Computers*, Vol.1, 2004, pp. 310–315.

[20] Guerci, J. R., R. M. Guerci, M. Ranagaswamy, et al., "CoFAR: Cognitive Fully Adaptive Radar," presented at the *IEEE Radar Conference*, 2014.

[21] Richards, M. A., *Fundamentals of Radar Signal Processing*, New York: McGraw-Hill, 2005.

[22] Skolnik, M. I. (ed.), *Radar Handbook*. New York: McGraw-Hill, 2008.

[23] Van Trees, H. L., *Optimum Array Processing: Part IV of Detection, Estimation, and Modulation Theory*. New York: Wiley Interscience, 2002.

[24] Farina, A., *Antenna-Based Signal Procesing for Radar Systems*, Norwood, MA: Artech House, 1992.

7

Advanced MIMO Analysis Techniques

All of the MIMO radar techniques covered in the book have been developed to improve radar system performance in challenging operating environments. In particular, the techniques presented in Chapters 5 and 6 were specifically developed to address the challenges encountered with traditional single waveform systems when trying to detect targets in real-world interference environments characterized by highly nonstationary heterogeneous ground clutter. Thus, it is critical to have realistic modeling and simulation tools available to fully develop and analyze these emerging optimal MIMO techniques. Further, since the techniques actually require adaptation of the waveforms on the fly it is not possible to simply use a static set of experimental data or even simulated data to analyze the algorithms. This poses a key challenge. This chapter describes an overall approach to modeling realistic radar clutter environments and shows how this approach has been applied successfully to analyze traditional systems and modifications that have been made to support the development of optimal and adaptive MIMO techniques.

7.1 Site-Specific Simulation Background

High-fidelity simulations of GMTI radar systems have proven extremely useful as a means to identify, analyze, and characterize real-world effects that significantly degrade radar detection performance such as heterogeneous clutter due to varying terrain and land cover and high densities of ground traffic. An example is ISL's site-specific radar simulator RFView [1] that was used with great success under the DARPA KASSPER program [2] to generate high-fidelity site-specific data sets that have been used by scores of researchers to develop and test new radar signal processing algorithms. Specifically the KASSPER Challenge Data Set has been reported on extensively in the literature (e.g., [3–53]).

We will begin by describing the basic modeling approach that includes the use of very accurate terrain and land cover databases to support the simulation of realistic radar clutter data. Typical sources of this data include

- *Terrain height information*: 1/3 arc second United States Geological Survey (USGS) National Elevation Dataset (NED) (~10m resolution) [54];
- *Roads and rivers*: U.S. Census Bureau TIGER/Line data [55];
- *Land cover type*: United States Geological Survey (USGS) National Land Cover Dataset (NLCD) (~30m resolution) [56].

Site-specific effects are significant, particularly in a mountainous environment. The inhomogeneities resulting from mountains and similar terrain will affect the performance of STAP and impact the training of the algorithms. Thus, site-specific phenomenology modeling is needed to properly capture these effects. The phenomenology modeling tool used to characterize the ground clutter is ISL's RFView software [1]. This tool has been used extensively in the characterization of system performance and STAP algorithm development and analysis (e.g., [41–43]). Use of this model allows for terrain-specific analyses of signal propagation and scattering. Modeling of ground scatter (i.e., clutter and hot clutter) is illustrated in Figure 7.1. For a given scenario, the signal environment produced by a transmitter is completely characterized in terms of the

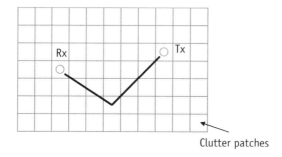

Clutter patches

Figure 7.1 The overall simulation approach is to break the world up into a large number of small clutter patches and use electromagnetic propagation and scattering models to characterize each clutter path between transmitter and receiver. (©ISL, Inc. 2018. Used with permission.)

signal strength, delay, Doppler, and AoA for each scattered signal or clutter patch.

The complex scattering amplitude of each clutter patch can be computed using either experimental data such as that found in [57] or using physics-based model such as a two scales of roughness model like the one first presented [58]. The main benefit of the two scales of roughness model is that it can model a much wider range of frequencies and geometries than empirical approaches including both monostatic and bistatic geometries. The two scales of roughness scattering model [58] is the summation of two cross-section terms: one accounting for the large scales of roughness (generally larger than the radar wavelength, that is, quasi-specular) and the other accounting for scales of roughness smaller than the radar wavelength (i.e., the Bragg-scatter contribution [58]). For terrain scattering, the large scale can generally be represented by the local terrain slope, while the shorter scale represents the local random roughness of the clutter patch. The terrain slopes are derived from the terrain data base and the parameters of the roughness model can be set based on the type of land cover derived from the land cover database. We note that the model is also a function of the electrical parameters of the clutter patch. These too can be selected using the land cover data. For example, if the clutter patch is water then appropriate conductivity and relative permittivity values would be chosen for the type water (i.e., freshwater versus saltwater).

The propagation from the platforms to each scattering patch is computed using an appropriate ray-tracing code. The calculation included the terrain database and also models of man-made structures such as buildings and towers. A simple but somewhat computationally intensive technique is to extract a terrain profile between the platform and clutter patch and use it to determine if there is line of sight to the patch. When there is line of sight the propagation coefficient is set to unity and zero otherwise. This model works well at higher frequencies. At lower frequencies where diffraction and multiplath effects are more pronounced, more sophisticated propagation models can be used (e.g., [59]).

Site-specific modeling begins with the selection of a geographic location in the world. For example, Figure 7.2 shows a scenario in Southern California characterized by mountainous terrain. This is the scenario that was used for the aforementioned KASSPER Challenge data set. Once the scenario is chosen the site-specific clutter can be computed by loading in all the terrain and land cover data for the desired location. Figure 7.3 shows the ground clutter computed using RFView for this scenario. The radar is located at the center of the scene and is flying at an altitude of 3 km above local terrain. We see that the resulting clutter power is highly variable due to the significant terrain relief in the scene. Along with the clutter power, the clutter signal delay, Doppler, and AoA are also characterized for each patch. These values are then used to both simulate realistic IQ data samples and space-time clutter covariance matrices for analyzing signal processing techniques.

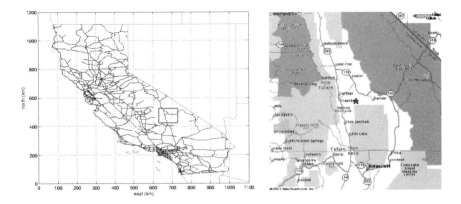

Figure 7.2   Site-specific radar analysis starts with selecting a geographic location of interest.

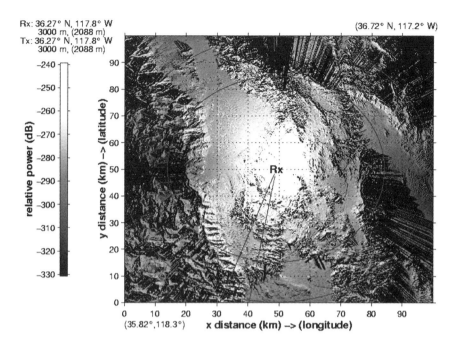

Figure 7.3 RFView clutter map. The radar is located at the center of the scene.

A general model for radar ground clutter si gnal is given as[1]

$$x(k,m,n) = \sum_{p=1}^{P_{cc}} a_p s_n\left(kT_s - \frac{r_{p,m,n}}{c}\right) e^{\frac{j2\pi r_{p,m,n}}{\lambda}} \mu_{p,m,n} \tag{7.1}$$

where k is the range bin index, m is the pulse index, n is the antenna index, $s_n(t)$ is the radar waveform, T_s is the sampling interval, λ is the radio wavelength, c is the speed of light, and $r_{p,m,n}$ and $\alpha_{p,m,n}$ are the two-way range and complex scattering amplitude, respectively, for the pth ground clutter patch on the mth pulse and nth antenna. The term $\mu_{p,m,n}$ allows for other random modulations of the radar data as needed in the pulse and channel dimensions. We see that this model results in the well-known radar datacube or 3-D data matrix. As formulated in (7.1) the datacube will be a matrix with dimensions range by pulse by channel. The waveform depends on the channel index, which allows us to use this model

---

1. The authors acknowledge Paul Techau for this notation and for his contributions to the development of radar signal models.

to simulate MIMO systems. We note that $\alpha_{p,m,n}$ will include any changes in the ground clutter reflectivity and transmit and receiver antenna response for a given clutter patch due to aspect angle changes resulting from motion of the platform during the CPI or motion of the clutter patch (e.g., wind-blown foliage [60, 61]). The latter is often called internal clutter motion (ICM).

We can often reduce computations for the general model given in (7.1) by assuming that the transmit and antenna responses and average ground clutter reflectivity for a given patch are constant during the CPI. Thus the model becomes

$$x(k,m,n) = \sum_{p=1}^{P_{cc}} a_p s_n \left(kT_s - \frac{r_{p,m,n}}{c} \right) e^{\frac{j2\pi r_{p,m,n}}{\lambda}} \mu_{p,m} \qquad (7.2)$$

where α_p now represents the complex ground reflectivity and transmit and antenna pattern for each patch (assumed constant throughout the CPI) and $\mu_{p,m}$ is a random modulation across the pulses for the pth patch due to ICM. Typically the realizations of $\mu_{p,m}$ are chosen to be Gaussian with a power spectral density consistent with the Billingsley spectral model [60] for ICM (see [61] for details about how $\mu_{p,m}$ is computed). The advantage of this form is that the ground clutter reflectivity does not need to be recomputed for each pulse (platform location) during the CPI. Since the modulation due to effects such as ICM will generally be more significant than modulations due to aspect angle changes (i.e., scintillation) this form of the model still retains a high level of fidelity. Also, it is important to note that this formulation of the model does include the effects of range and Doppler migration of the scatterers due to the motion of the platform during the CPI, which will be an important effect to capture for long CPIs. These effects can be readily included; however, they require more computing resources to simulate the data. The values chosen for α_p are drawn from zero-mean complex Gaussian distributions with variance equal to the average scattered power of each patch as computed using the scattering models discussed above. We will show below, however, that the resulting distribution of the simulated radar data can deviate significantly from Gaussian depending on the terrain type. This model has been used extensively to simulate realistic airborne radar clutter data. A typical example was given in Figure 3.14. More examples are shown later in this chapter.

We also often wish to have an accurate model of the clutter co-variance matrix. The typical monostatic clutter environment may be approximated by a large number of typically stationary scatterers. If we ignore range sidelobes, the scatterers that contribute to the clutter return in a given range bin r_o will reside in an annulus of inner radius $r_o - \Delta r/2$ and outer radius $r_o + \Delta r/2$, where Δr is the radar range resolution.

We assume that the radar has N elements and M pulses. The scattered energy received in a given range bin is determined from the sum over all the scatter from that range bin. As stated above, it may be approximated by a large number of individual scatterers. These scatterers must be characterized in terms of: range from the platform, the AoA, scatter amplitude, and Doppler shift. The scatterers may be patches of ground (i.e., area scatter) or point scatterers such as clutter discretes. The clutter for a given range bin is the sum over all of these scatterers. Thus the clutter samples from the kth range bin are given as

$$\mathbf{x}_k = \sum_{p=1}^{P_{cc}} \alpha_p \mathbf{v}(\theta_p, f_p) \in C^{MN} \tag{7.3}$$

where $\mathbf{v}(\theta_p, f_p)$ is the space-time steering vector for AoA θ_p and Doppler f_p, P_{cc} is the number of scatterers in range bin k, and in the case of area scatter, the α_p are complex, independent, Gaussian random numbers chosen to have a variance that corresponds to the powers predicted by RFView. Here we are ignoring the effects of nonzero bandwidth on the array response.

The space-time steering vector has the form

$$\mathbf{v}(\theta_p, f_p) = \mathbf{b}(f_p) \otimes \mathbf{a}(\theta_p) \tag{7.4}$$

where $\otimes$ denotes the Kronecker or tensor matrix product, $\mathbf{a}(\theta_p)$ is the array response at AoA θ_p, and $\mathbf{b}(f_p)$ is the temporal steering vector for Doppler frequency f_p. These vectors may be represented by

$$\mathbf{a}(\theta_p) = \begin{bmatrix} 1 & e^{j\phi(\theta_p)} & \ldots & e^{j(N-1)\phi(\theta_p)} \end{bmatrix}'$$

$$\mathbf{b}(f_p) = \begin{bmatrix} 1 & e^{j2\pi f_p T_r} & \ldots & e^{j(M-1)2\pi f_p T_r} \end{bmatrix}' \tag{7.5}$$

where we have assumed a uniform linear array, $\phi(\theta_p)$ is the phase shift relative to element #1 for a signal arriving at AoA θ_p, and T_r is the PRI (equal to the reciprocal of the PRF). Thus each range sample $\mathbf{x}_k$ is an $NM\times1$ vector that represents a snapshot of the N antenna array elements for each of the M pulses.

The ideal ground clutter covariance matrix is computed as $E\{\mathbf{x}_k\mathbf{x}'_k\}$[62] and is given as follows:

$$R_{cc} = \sum_{p=1}^{P_{cc}} |\alpha_p|^2 \mathbf{v}(\theta_p, f_p)\mathbf{v}'(\theta_p, f_p) \tag{7.6}$$

where H is the Hermitian transpose operator. This covariance model can be readily used to analyze MIMO systems by simply replacing $\mathbf{v}(\theta_p, f_p)$ with the MIMO version discussed in Chapter 3. We note that this covariance model can also be augmented to include fast-time (range) DoF to account to support the analysis of joint clutter and terrain-scattered interference (TSI) mitigation [62].

The signal model in (7.2) and covariance model in (7.6) represent a baseline simulation capability with broad applicability. These models can be readily augmented to account for other important effects as needed to support more advanced analysis. For example, a method to incorporate losses due to ICM caused by wind-blow foliage is given in [61]. A method for using these models to analyze the impact of ground target signals in the clutter training data is given in [63, 64]. Also, these models can be used to simulate bistatic systems and with an appropriate model of the antenna can also be used to model systems with polarimetric sensing antennas. Further, antenna calibration and receiver and transmitter channel errors can be included by again making appropriate modifications to the antenna model. For example, antenna position errors can be modeled in (7.2) by simply modifying the value of $d_{p,n,m}$ for each antenna.

While most of the results in this chapter focus on using the simulations to analyze generally narrowband GMTI-type radars, we note that this model has been used in the past to simulate much wider bandwidth system, including SAR systems. The main signal effect that differentiates SAR from GMTI is the significant degree of range walk during the radar CPI. Since (7.1) allows for a unique time delay on each pulse, the model captures these effects and therefore can produce realistic SAR data. As an example, Figure 7.4 shows a simulated SAR data set with two tank targets, each composed of a large number of point scatterers. In this case, each clutter α_p in (7.1) is replaced with the complex scattering for the target model. The SAR processing compensates for the radar motion during the CPI for a single point at the center of one of the targets. Scatterers away from this point are blurred because of the uncompensated motion. So we see that one target is focused while the other is blurred. This data set highlights the fidelity of the simulated data.

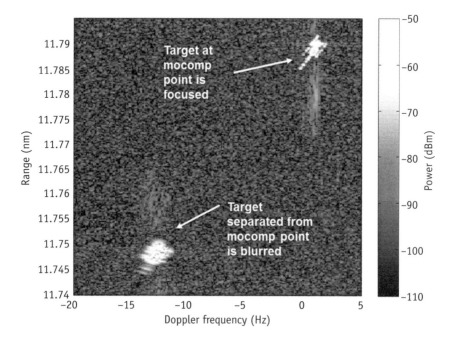

Figure 7.4 Simulated SAR data for two simple tank models. The data is motion compensated to the location of the target in the upper right hand corner.

7.2   Adaptive Radar Simulation Results

We highlight the importance of the site-specific approach by showing some key results from processing KASSPER data set 2 [65]. These results were previously presented in [66]. A list of system parameters are shown in the table in Figure 7.5. This data set simulates an X-band radar and includes site-specific clutter computed using DTED Level 1, and therefore this data set represents a generally heterogeneous clutter environment. The simulated system has 38 pulses and 12 spatial channels. Additional details about the data set and system parameters can be found in [65].

Figure 7.6 shows the SINR loss [67] for multibin (three adjacent bins) post-Doppler channel-space STAP filter weights computed using the ideal covariance matrix for each range bin. SINR loss is the ratio of the SINR to the SNR that would be observed in a noise-limited environment [67]. The mainbeam clutter notch is located at the two-way Doppler of −12 m/s due to the antenna being steering away from broadside. We see that the mainbeam clutter notch width exhibits significant variability due to terrain-induced heterogeneity. At some ranges there is no loss because the clutter is shadowed by terrain.

Figure 7.6 also shows the SINR loss surface for a sample matrix inverse (SMI) multibin post-Doppler channel space reduced-DoF

Parameter	Value
Carrier frequency	1240 MHz
Bandwidth	10 MHz
Number of pulses	32
Minimum range (one way)	35,000 meters
Maximum range (one way)	50,000 meters
Pulse repetition frequency	1984 Hz
Peak power	15 kW
Duty factor	10%
Noise figure	5 dB
System losses	9 dB
Antenna	8 vertical by 11 horizontal broadbeam elements
Crab	3 degrees
Front-to-back ratio (two way)	25 dB
Platform azimuth heading	270 degrees
Platform height above local terrain	3000 meters
Platform speed	100 m/s

Figure 7.5   KASSPER data set 2 parameters.

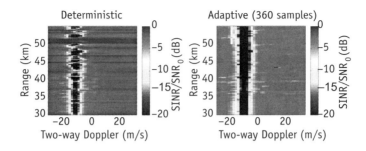

Figure 7.6 Example of processed site-specific simulated radar data. Left: ideal clutter filter. Right: adaptive clutter fitler. (©2004 IEEE. Reprinted, with permission, from *Proceedings of the 2004 IEEE Radar Conference.*)

STAP algorithm. Again, three Doppler bins are used so that the number of adaptive DoF is 24. In this case a relatively large training window of 360 samples (5.4 km) was used to estimate the sample covariance matrix. Thus we see that the variability of the clutter notch observed in the ideal covariance case is reduced due to the averaging process and in general the notch width over the entire range dimension is wider. Comparing the two results in Figure 7.6 we conclude that localized processing strategies will generally lead to improved sensitivity in detecting targets with low radial velocities and lower system MDV.

Clearly, the impact of site-specific terrain is evident from this example. Simulations involving a bald earth assumption would lead to very different conclusions about the radar performance and also to STAP training algorithms that likely would not work well in the real world. One of the most interesting aspects of the site-specific analysis approach is that it can be used to analyze performance for any location in the world and help radar designers understand how varying clutter scenes can impact performance. Next we highlight this capability by showing a number of clutter simulations for different geographic locations and show how the clutter statistics can varying significantly from site to site.

7.3   Site-Specific Clutter Variability and Statistics

Nine simulation scenarios were chosen in the continental United States spanning various terrain types. The locations of the scenarios are shown in Figure 7.7. High resolution (10m grid spacing)

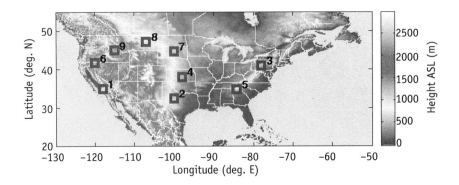

Figure 7.7   Locations of the nine site-specific simulation scenarios.

terrain data from the United States Geological Survey's (USGS) National Elevation Dataset (NED) was used for the simulations. The terrain for each location is shown in Figure 7.8. The area number listed above each plot corresponds to the numbers marking the scenarios shown in Figure 7.7. Table 7.1 lists the standard deviation of the terrain heights for each scenario.

ISL's RFView model was used to compute a site-specific clutter map for each of the scenarios. In each case the radar platform was located directly to the east at a ground range of approximately 50 km from the center of the scene. The radar altitude is 10 km, the heading is north, the speed is 150 m/s, and the operating frequency 10 GHz.

The simulated clutter maps for each scenario are shown in Figure 7.9. The clutter patch size is 5m × 5m. We see that the simulated clutter varies significantly among the nine scenarios. As expected, the scenarios involving relatively flat terrain result in clutter maps with little variation over the scene, while the mountainous scenarios result in significant variations in the clutter power as well as deep shadow regions. We will see below that the variability of the clutter power in the rough terrain scenarios will lead to clutter distributions with heavier tails relative to the clutter distributions for the flat terrain scenarios.

The clutter patches were sorted into radar range-Doppler-azimuth resolution cells to form a representation of the radar data cube. The range-Doppler clutter maps for a single azimuth angle are shown in Figure 7.10 for each simulation scenario. The range resolution was assumed to be 15m (10-MHz bandwidth), the azimuth resolution was assumed to be 1° (1.7m antenna aperture),

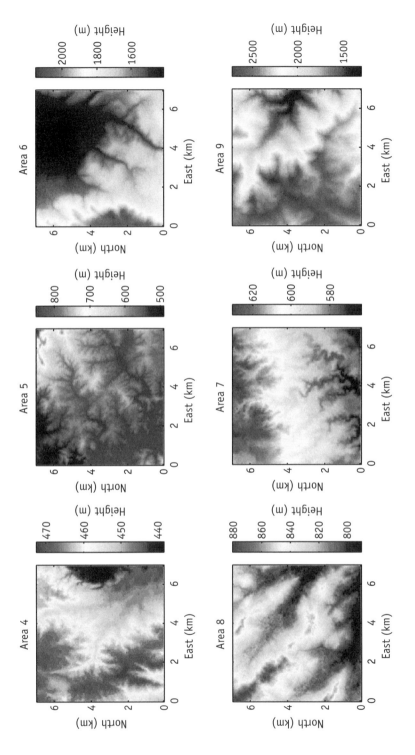

Figure 7.8 Terrain height maps of the nine site-specific simulation scenarios show in Figure 7.7.

Table 7.1
Estimated Terrain Height Standard Deviation for Each Simulation Scenario

Scenario Number								
1	2	3	4	5	6	7	8	9
std (m) 28.7	5.0	57.2	8.1	41.0	162.5	15.3	17.2	308.3

and the Doppler resolution was assumed to 1 m/s (30-ms coherent processing interval). All the patches in a given resolution cell were found and their power as predicted by RFView based on the ground clutter RCS model was summed to form the radar data cube. It is important to note that for this analysis it was the sum of the power predicted by the site-specific clutter model that was used in the data cube characterization and not the sum of the realizations of the complex voltage as discussed below. We note that this representation of the radar data cube is the ensemble average power of the data cubes simulated by sorting the random realizations of the complex clutter voltages for each patch when the random voltages are drawn from independent distributions. It will be shown below that the sum of the power will be useful when deriving a distribution for the clutter.

As discussed above, site-specific simulations usually assume a particular distribution for each clutter patch for use in arriving at a realization of the complex scattering voltage. We typically assume that the voltage for each patch is drawn from a zero-mean complex Gaussian distribution with variance equal to the scattering power predicted by a site-specific model such as RFView. We are interested in deriving a distribution of the resulting set of random numbers. To this end we begin by defining the cumulative distribution function (CDF) of the random variable that represents the set of clutter patch voltages:

$$F_Z(z) = \text{Prob}(Z < z) \tag{7.7}$$

where the operator $\text{Prob}(\bullet)$ is the probability of the event in parentheses. The probability is found by simply adding up all the probability associated with getting a value of $Z < z$ when a member is drawn from the set, which is given as

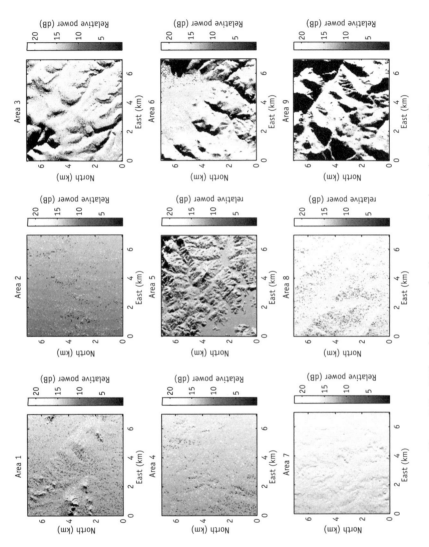

Figure 7.9 RFView clutter maps for the scenarios shown in Figure 7.7.

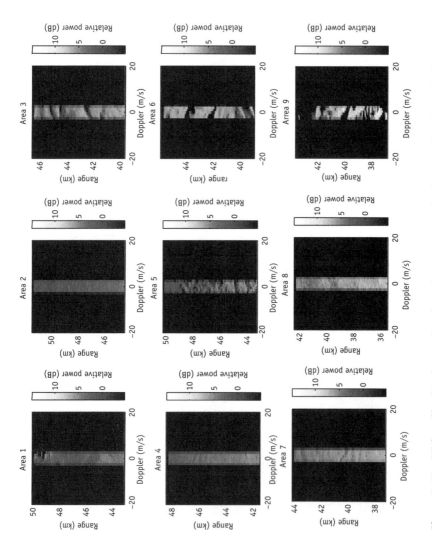

Figure 7.10 RFView Simulated range-Doppler radar power maps for the scenarios shown in Figure 7.7.

$$F_Z(z) = \sum_{p=1}^{P} \rho_p \mathrm{Prob}(Z_p < z) = \sum_{p=1}^{P} \rho_p F_{Z_p}(z) \tag{7.8}$$

where P is the total number of clutter patches, ρ_p is the probability of drawing the pth element, and $F_{Z_p}(z)$ is the CDF of the pth element in the set. The pdf of Z is readily found by taking the derivative of (7.8) with respect to z, which gives

$$f_Z(z) = \sum_{p=1}^{P} \rho_p f_{Z_p}(z) \tag{7.9}$$

where $F_{Z_p}(z)$ is the pdf of the pth clutter patch. As discussed above, we will assume that the clutter patch probability densities are independent zero-mean complex Gaussian conditioned on the specific clutter power of the patch[2]. Thus, the clutter patch distribution will be independent zero-mean conditional Gaussian with variance equal to σ_p^2, the power of the pth patch as predicted by the site-specific clutter model. Since for radar performance prediction we will generally be interested in computing the false-alarm probabilities at the output of an amplitude or power threshold detector, we will focus on the distributions of the clutter patch amplitudes, which will be conditionally Rayleigh distributed [68] in this case and given as

$$f_{Z_p|\sigma_p}(z|\sigma_p) = \frac{2z}{\sigma_p^2} e^{-z^2/\sigma_p^2} \tag{7.10}$$

where the $\alpha|\beta$ notation indicates a pdf of α conditioned on the value of β and we note that this density function is zero for $z < 0$. Now, if we draw the clutter patches with equally likely probability (i.e., $\rho_p = 1/P$), then the distribution of the collection of clutter patch amplitudes will be[3]

---

2. The authors acknowledge Paul Techau for this interpretation of the clutter distribution.
3. The authors acknowledge Dr. Christopher Teixeira and Paul Techau for originally suggesting this solution.

$$f_{Z|\sigma_1,\sigma_2,\dots,\sigma_P}\left(z\,\middle|\,\sigma_1,\sigma_2,\dots,\sigma_P\right)=\frac{1}{P}\sum_{p=1}^{P}\frac{2z}{\sigma_p^2}e^{-z^2/\sigma_p^2} \qquad (7.11)$$

Again we note that this density function is zero for $z < 0$. A similar expression was used to analyze the distribution of SAR image histograms in [69]. If we were to treat the clutter patch powers (σ_p) as random variables then a potential approach would be to determine their joint probability density functions for various scenarios using a site-specific model and then compute the clutter patch marginal density function $f_Z(z)$ by integrating over all possible values of σ_p. That is, we could compute

$$f_Z(z)=\int_{-\infty}^{\infty}\int_{-\infty}^{\infty}\cdots\int_{-\infty}^{\infty}f_{Z|\sigma_1,\sigma_2,\dots,\sigma_P}\left(z\,\middle|\,\sigma_1,\sigma_2,\dots,\sigma_P\right)$$

$$f_{\Sigma_1,\Sigma_2,\dots,\Sigma_P}(\sigma_1,\sigma_2,\dots,\sigma_P)d\sigma_1,\sigma_2,\dots,\sigma_P \qquad (7.12)$$

The approach we will take here, however, is to simply use the computed values for the patch powers for a given scenario and use them to compute the exact *site-specific* clutter density for the particular given scenario of interest using (7.11). We note that since our ultimate goal is to predict site-specific radar performance it is appropriate to work with a pdf conditioned on the information (in this case the clutter patch powers) about a particular scenario of interest.

The result in (7.11) gives the density for the raw clutter patches (see Figure 7.9); however, we ultimately are interested in determining the density of the clutter as observed by the radar. Thus, we are interested in deriving a density for a collection of radar resolution cell outputs that we will represent with a random variable Y. For simplicity we will assume an ideal radar that simply sums the contributions from all the clutter patches within a given radar resolution cell. For the clutter model discussed above, which assigns a random realization of a zero-mean complex Gaussian voltage to each clutter patch, the output of the radar resolution cell will also be zero-mean Gaussian (conditioned on the site-specific clutter patch powers) since it is just the sum of a number of independent Gaussian random variables. The variance of the output will be the sum of the variance of the input clutter patches or simply the sum of the power of each patch. Thus, we see that the data cube simulation that sums the power for the clutter patches in each cell represents the parameter needed to characterize the clutter distribution for each cell. Based on this argument and following the same approach used above, the conditional pdf of the collection of radar resolution cells will be

$$f_{Y|r_1,r_2,\ldots,r_{N_c}}(y|r_1,r_2,\ldots,r_{N_c}) = \frac{1}{N_c}\sum_{n=1}^{N_c}\frac{2y}{r_n}e^{-y^2/r_n} \tag{7.13}$$

where r_n is the sum of the power of the clutter patches contributing to the nth radar resolution cell and N_c is the total number of resolution cells being analyzed. We see that this distribution is identical in form to the distribution of the raw clutter patches given in (7.11). Also, we note that since the parameters r_n are functions of the clutter patch powers that the density function given in (7.13) is also conditioned on the parameters σ_p.

The distributions of the raw clutter patches computed using (7.11) for each of the scenarios discussed above are shown in Figure 7.11. To facilitate comparisons of the resulting distributions with the baseline Rayleigh distribution, the clutter maps were also generated for a sandpaper Earth (i.e., all the terrain values in a given scenario set equal to the mean of the corresponding terrain map shown in Figure 7.9). The clutter maps for the terrain and sandpaper Earth cases were then both normalized by the average value of the power of the sandpaper Earth patches. Thus if the power of the patches for the sandpaper Earth case are all equal, the final distribution given by (7.11) will be Rayleigh with the parameter equal to unity. Therefore, we expect the distribution of the sandpaper Earth case to be approximately equal to the Rayleigh distribution with only a small deviation due to the small variations in the clutter patch powers over the scene due to small slant range and grazing angle differences.

The distributions in Figure 7.11 reveal the expected behavior in that the sandpaper Earth model results in clutter distributions that are nearly identical to Rayleigh and that for the smoother terrain scenarios the resulting clutter distributions are also very close to Rayleigh. The rough terrain cases on the other hand result in distributions that deviate significantly from the Rayleigh model

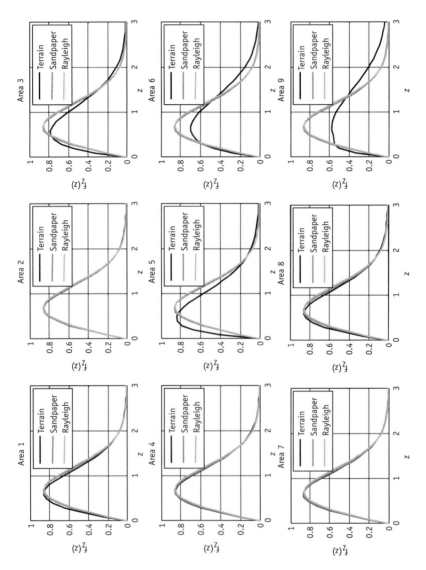

Figure 7.11   Clutter patch pdfs for the scenarios shown in Figure 7.7. The area number is given in the title.

(we note that the distinction between smooth and rough terrain as used here is a qualitative assessment based on the observed differences in the actual terrain and resulting clutter maps). In general these distributions exhibit much heavier tails than the Rayleigh case, which will lead to higher radar false-alarm rates. Figure 7.12 shows the same distributions given in Figure 7.11 on a decibel (dB) scale, which helps demonstrate the heavier tails exhibited by the rough terrain cases.

The distributions of the radar data were also computed using (7.13) for the range-Doppler radar clutter maps shown in Figure 7.10. As with the raw clutter patches the distributions were also computed for the sandpaper Earth clutter model and scaled as discussed above, so that for the case when the clutter power in all the resolution cells is equal, the distribution will exactly match the Rayleigh distribution. Also, only nonzero resolution cells were included in the calculation (i.e., radar resolution cells blocked by terrain shadows or in Doppler bins outside the azimuth beam of interest were omitted). The resulting distributions are shown on a linear scale in Figure 7.13 and on a dB scale in Figure 7.14 to emphasize the heavier tails for the rough terrain cases. We see similar behavior as we did with the raw clutter patches, however in this case the rough terrain cases reveal an ever greater deviation from the Rayleigh distribution. We also note that the apparent discontinuity in the curve for simulation 9 seen in Figure 7.13 at approximately $y = 0.05$ is an artifact of the interval used to produce the plot

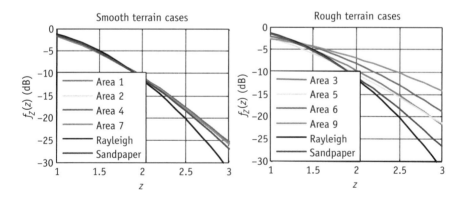

Figure 7.12 Clutter distributions for the raw clutter patches plotted on a decibel scale. Left: Smooth terrain cases. Right: Rough terrain cases.

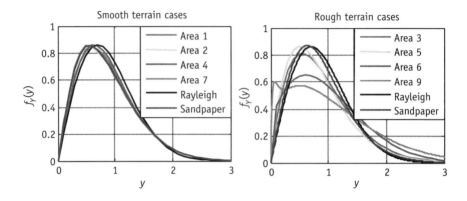

Figure 7.13 Clutter distributions for the radar data cube. The azimuth angle is 270°. Left: Smooth terrain cases. Right: Rough terrain cases.

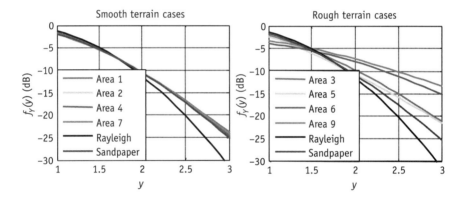

Figure 7.14 Clutter distributions for the radar data cube plotted on a decibel scale. The azimuth angle is 270°. Left: Smooth terrain cases. Right: Rough terrain cases.

and that the curves are indeed smooth and integrable for all values of $y > 0$.

7.4 Optimal Waveform Analysis

The previous sections in the chapter showed how site-specific simulations can be used to generate realistic data for radar analysis. The ultimate goal is to use them to support the development of optimal MIMO techniques. In this section we present an example of how these tools can be used to analyze the performance of an adaptive transmitter [70–73]. An adaptive transmitter is a wave-form optimization technique that can be used to adjust any or all of the radar transmitter DoF based on observations of a dynamic environment caused by intentional and/or unintentional interference sources. In this example we will adapt the transmitter waveform to improve performance against terrain-scattered interference caused when transmitter energy is scattered off terrain features such as mountains and enters into the antenna mainbeam. This phenomenon is sometimes referred to as hot clutter. This scenario was first presented in [74].

The simulation scenario considered is shown in Figure 7.15. The scenario involves a stationary jammer (Tx) located on a mountain peak and an airborne radar (Rx) traveling east at 150 m/s. The simulated radar system is a UHF radar with 1-MHz receiver bandwidth. The intersection of the simulated antenna beampattern with the ground is shown in Figure 7.15. Figure 7.16 shows the simulated TSI caused by the interference. The TSI is shown for both an omnidirectional antenna and for the true simulated antenna pattern. We see that even though the interference is outside the radar mainbeam, significant mainbeam interference energy is still present due to the terrain. These clutter maps were generated using ISL's RFView software, which uses the simulation methodology discussed above. We note that in this case the simulation of the clutter is for a bistatic geometry.

For this simulation it was assumed that the direct path interference-to-noise ratio is 60 dB, which is based on the following system

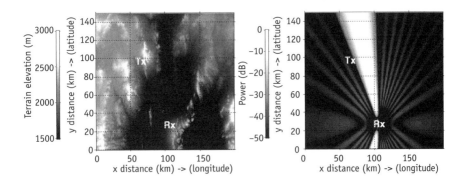

Figure 7.15 Terrain-scattered interference scenario. Left: terrain map with transmitter ('Tx') and receiver locations ('Rx'). Right: receiver antenna patterns.

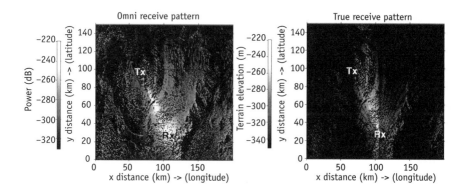

Figure 7.16 Terrain-scattered interference scenario. Left: Terrain-scattering interference (omni receiver antenna pattern). Right: Terrain-scattered interference with receiver antenna pattern included.

parameters: radar antenna gain of 20 dB, interference transmit gain of 5 dB, 100-W interference transmit power, radar receiver noise factor of 5 dB, and noise temperature of 290K. The simulation assumes that the interference direct path signal is insignificant relative to the TSI due to the steering of the array. A temporal interference covariance matrix as observed at the radar was simulated using the approach given in [62]. The covariance matrix was computed for 100 time samples separated by the inverse system bandwidth. The eigenvalues of the interference covariance matrix are shown in Figure 7.17. We see that there is a large spread in the power of the eigenvalues of the interference signal. This is significant because adaptive transmitter theory predicts that the maximum gain of an

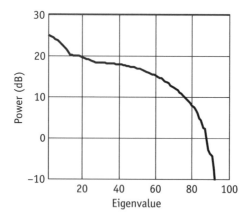

Figure 7.17 Eigenvalues of the Terrain-scattered interference covariance matrix.

adaptive transmitter relative to any arbitrary waveform is limited to the power spread in the interference eigenvalues [73, 75]. Thus, based on the eigenvalues shown in Figure 7.17 there is potential to significantly improve the system sensitivity by adapting the transmit waveform.

As discussed in Chapter 5, adaptive transmitter theory states that the waveform that optimizes output SINR is the eigenfunction (eigenvector for discrete time analysis) associated with the smallest eigenvalue of the interference covariance function [71]. While the optimal waveform maximizes SINR, it often results in waveforms that are not practical for use in radar systems because the important properties such as low-range sidelobes and high-range resolution are lost in the optimization. An approach that allows the waveform optimization to account for both SINR and a desired radar waveform is given in [72]. The techniques use a specified number of the interference eigenvectors associated with the smallest eigenvalues as opposed to a single eigenvector. This relaxes the optimization and provides DoF to match the waveform in a least-squares sense to a desired waveform such as a traditional linear FM chirp (i.e., LFM). In this manner, new optimization techniques allow one to trade SINR for more desirable waveform properties.

Here we will consider two metrics presented in [72] to characterize the performance gains of the adaptive transmitter. The first is the ratio of the adaptive transmitter waveform SINR to the SINR of a traditional LFM waveform at the output of a standard receiver filter matched to the input waveform. We denote this metric as $SINR_o/SINR_{LFM}$. The second metric is the same ratio but computed at the output of an optimal receiver whitening filter [75]. This metric is denoted as $SINR_{0w}/SINRL_{FMw}$. Figure 7.18 shows the metrics for the scenario discussed above as a function of the number of eigenvalues used in the optimization. We see that the performance gains at the output of the whitening filter are smaller relative to the standard matched filter, which is expected since receive-only whitening filters are known to improve SINR for hot clutter [62]. We also see that the gain of the adaptive transmitter decreases as a larger number of the interference eigenvalues are used in the optimization. This is expected because more eigenvectors are used the correlation between the transmitted signal and therefore the interference will increase, resulting in lower SINR.

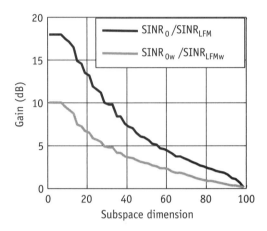

Figure 7.18   Performance gain of the adaptive transmitter relative to a conventional LFM waveform.

Figure 7.19 shows the compressed waveforms for several values of the number of eigenvectors used in the optimization. We see that the range resolution and range sidelobe levels generally improve as the number of eigenvectors is increased and the resulting waveform provides a better match to the desired LFM waveform. The results are shown both with and without the receiver whitening filter. As can be seen the whitening filter negatively impacts the range sidelobe level. The LFM waveform with a receiver whitening filter is shown for comparison. This result is provided to illustrate that using receive-only processing (i.e., no adaptivity on transmit) still results in significant degradation to the radar waveform.

This example simulation shows how advanced waveform optimization techniques can improve system performance when an accurate model of the interference environment is available. In this case, by including realistic terrain-scattered interference, we were able to take advantage of the colored interference spectrum to develop waveforms that perform better than those designed without any detailed knowledge of the radar operating environment. In practice we would expect the radar to adapt the waveform on the fly based on the observed interference statistics or channel model as discussed in Chapter 6. We will show next that this requires a simulation approach that allows the site-specific simulation tools discussed in this chapter to be included in the loop with the waveform optimization algorithms during the algorithm development and testing phases.

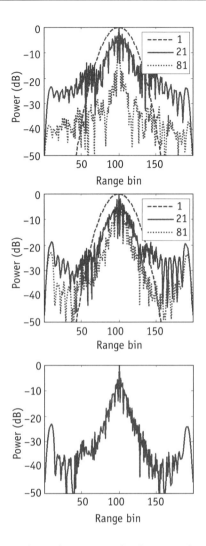

Figure 7.19 Compressed waveforms. Top: adaptive transmitter without whitening filter on receive. Center: adaptive transmitter with whitening filter on receive. Bottom: LFM waveform with whitening filter on receive.

7.5   Optimal MIMO Radar Analysis

The optimal MIMO techniques presented Chapters 5 and 6 sense, learn, and adapt (SLA) to complex environments. In general these approaches represent a *fully* adaptive solution that jointly optimizes the adaptive transmit and receive functions [76–79]. While modeling and estimation of complex ground clutter for airborne radar in support of STAP remains a challenge as discussed above, the problem is even more complex when attempting to jointly optimize both the transmit and receive functions. This is because the popular clutter covariance modeling approach discussed above is generally a nonlinear function of the space-time transmit signal. Nonetheless the site-specific, high-fidelity simulations are critical for predicting real-world performance. This is especially true when simulations are used as a substitute for actual flight tests as we are proposing.

As discussed earlier in this chapter, simulations with sufficient fidelity to support meaningful optimal waveform analysis involve very sophisticated electromagnetic propagation and scattering calculations based on terrain and land cover models. The data can often take several hours to generate, which can be prohibitive when attempting to generate data to test systems for which the waveform is continually changing, requiring rapid updates to the simulated or experimental test data.

Thus a simulation environment used to test the optimal MIMO algorithms will need to be capable of updating the simulated IQ data many times. This is in contrast to testing a system with receiver-only adaptivity where we can simulate the data once and use it to test out many different algorithm alternatives. The KASSPER Challenge data set discussed earlier in this chapter is a good example! Fortunately, for many situations, even though the optimal waveform is being updated, the underlying scenario or channel is not changing rapidly. We will show that this can be exploited to develop simulations that are highly efficient.

A new approach to signal-dependent clutter modeling was introduced in [80]. Unlike traditional signal-dependent stochastic models, a new MIMO stochastic transfer function approach was developed that results from a fundamental physics scattering model; that is, the Green's function. A key advantage of this new approach is that the transfer function is signal independent (i.e., transmit signal independent)—although of course the output (received) resultant clutter certainly depends on the chosen input (transmit function). This approach facilitates the joint optimization of the transmit space-time waveforms and receiver functions—a problem that is generally nonlinear for all but the simplest clutter models using traditional approaches [80, 81]. Moreover it greatly facilitates efficient modeling and simulation of optimal MIMO radar for which the transmit function is adaptive. Next we present show how the Green's function methodology can be implemented for simulating high-fidelity radar following the approach given in [82].

An existing and universally employed numerical approach for modeling the space-time GMTI (or AMTI) ground clutter signal is given by the Riemann sum approximation like the one presented earlier in this chapter:

$$\mathbf{x}_c(t) = \sum_{p=1}^{N_p} \alpha_p \tilde{\mathbf{w}}_p(t) \tag{7.14}$$

where α_p and $\tilde{\mathbf{w}}_p(t)$ are the complex amplitude and space-time radar wave (appropriately delayed and Doppler shifted) for the pth clutter patch in the scene, respectively. We note that $\mathbf{x}_c(t)$ and $\tilde{\mathbf{w}}_p(t)$ are space-time vectors with dimensions equal to the product of the number of spatial channels and number of radar pulses. We note that (7.14) is simply a different way of expressing the signal model give earlier in (7.2) in vector format that will help facilitate the analysis below. To simulate clutter data, $\mathbf{x}_c(t)$, this summation is typically performed over millions of patches N_p. The problem is exacerbated for optimal MIMO radar analysis since we have to recompute the sum every time a new waveform is estimated after each new optimization of the transmit waveform and receive beamforming weights.

Alternatively, if we model the clutter signal as the following sum:

$$\mathbf{x}_c(t) = \sum_{p=1}^{N_p} \mathbf{w}(t) * \underline{H}_p(t) \tag{7.15}$$

where each element of $\underline{H}_p(t)$ contains the clutter channel response or Green's function for a single radar channel and pulse. The operator * is the element-wise linear convolution operation. We see that in this case the radar waveform $\mathbf{w}(t)$ can be factored out of the sum as follows:

$$\mathbf{x}_c(t) = \mathbf{w}(t) * \sum_{p=1}^{N_p} \underline{H}_p(t) = \mathbf{w}(t) * \underline{H}_c(t) \qquad (7.16)$$

allowing us to compute the sum once to get the clutter channel and then recompute $\mathbf{x}_c(t)$ efficiently whenever $\mathbf{w}(t)$ is updated by using the much less computationally intensive convolution operation of the new waveform with the clutter channel. We note that this formulation assumes that higher-order Doppler effects (e.g., time dilation) are negligible during a single pulse, which is typically the case with most practical ground-based and airborne pulsed-Doppler radar systems.

The clutter channel matrix will depend on the site-specific clutter propagation and scattering as well as the nature of the transmit and receive hardware. Although our method would support any bandlimited model for the radar transmit and receiver chains, we assume a radar with an ideal (e.g., boxcar) transmitter and receiver frequency response. The received signal from a clutter patch for a single antenna and pulse is a function of the radar waveform, the transmitter channel response, the clutter patch response, and the receiver channel response as follows:

$$x_c(t) = w(t) * h_t(t) * h_s(t) * h_r(t)$$

where $w(t)$ is the radar waveform used to probe the channel, $h_t(t)$ is the transmitter channel response, $h_s(t)$ is the response of the clutter patch, $h_r(t)$ is the response of the radar receiver, and $*$ is the convolution operator. In order to compute the clutter channel (impulse response) we will assume that the radar waveform $w(t)$ is an ideal impulse. In this case $y_c(t) = h_c(t)$ the clutter channel response or Green's function. We will also assume that $h_t(t) = h_r(t)$ and that the clutter patch response is broad in frequency relative to the bandwidth of the radar and therefore can also be modeled as an ideal impulse response. Given these assumptions, the clutter channel response is equivalent to the response of the radar receiver, which can be expressed as the following inverse Fourier transform as follows:

$$h_c(t) = h_r(t) = \int_{-\frac{B}{2}}^{\frac{B}{2}} k e^{j2\pi ft} df = Bk\,\text{sinc}(\pi Bt)$$

where B is the system bandwidth and k is a constant that is the product of the complex amplitudes of the clutter patch, receiver, and transmitter channels. These amplitudes are computed based on the assumed gain and noise figure of the radar and the complex propagation environment due to terrain and clutter patch properties (land cover type). The advantage of this impulse response model for clutter is that it can be used to determine the input-output response of any other finite energy bandlimited transmit waveform. In this way, a fully adaptive radar that changes its waveforms on the fly can be accommodated.

Figure 7.20 demonstrates the equivalence of the efficient Green's function representation versus the traditional brute force approach of summing the transmit waveform return for every scatterer in the scene. On the top is a simulation using and LFM waveform and summing the waveform response for every clutter patch in the scene (traditional approach). The bottom plot shows the convolution of the Green's function with the LFM—which is to within numerical round-off equivalent to the first plot. Note that the top plot has a small thermal noise floor while the bottom plot is clutter only, and the top plot takes minutes to compute whereas the bottom plot takes on the order of one second. Thus, once the channel is computed the waveform can be changed without significant computational cost.

Figure 7.21 shows an example where the channel model was used to generate data for two different waveform types. Here we generate the radar data for an LFM waveform and for a random phase coded waveform using the same channel model. We can clearly see the differences in the processed radar data resulting from the two different waveforms. In particular, the random phase coded waveform exhibits much higher-range sidelobes. As with the previous example, then calculation of these two data sets is very efficient once the channel model is known whereas recomputing the simulated data for each new waveform assumption can be very time consuming.

The new Green's function approach to simulating complex environments supports very advanced analysis of optimal waveform

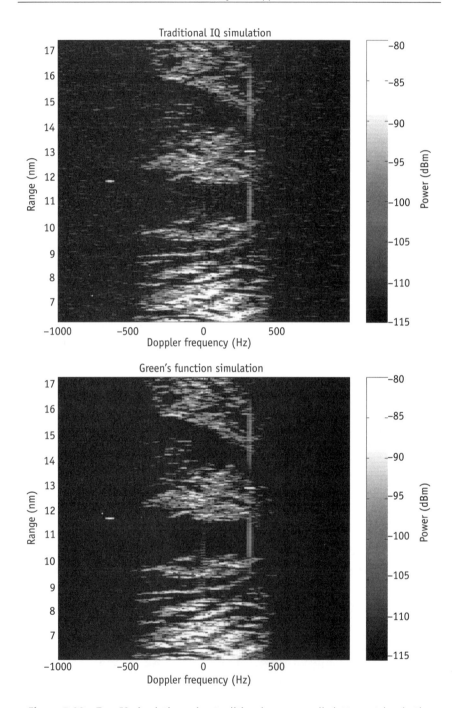

Figure 7.20 Top: IQ simulation using traditional sum over all clutter patches in the scene. Bottom: IQ simulation using the convoluation of the waveform with the clutter channel Green's function. (©2017 IEEE. Reprinted, with permission, from *Proceedings of the 2017 IEEE Radar Conference.*)

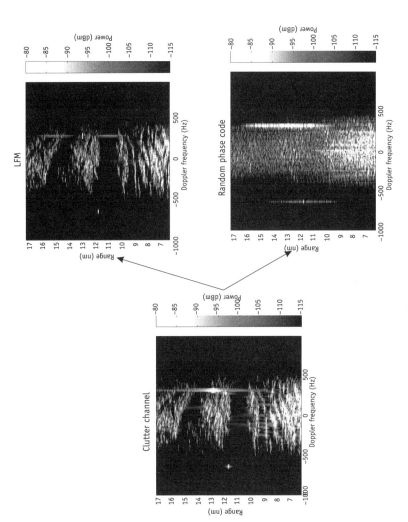

Figure 7.21 Top: IQ simulation using traditional sum over all clutter patches in the scene. Bottom: IQ simulation using the convolution of the waveform with the clutter channel Green's function.

systems. We demonstrate this capability with a final example involving a site-specific scenario with a radar flying along the coast
of Southern California. The scenario is shown in Figure 7.22. The
red line shows the airborne radar's trajectory. The grayed rectan-
gular area shows the region being surveilled by the radar.

Figure 7.23 shows the resulting clutter maps as a function of radar position (units are km). Note the presence of significant temporal and spatial variation—a prerequisite for optimal transmit radar
improvements.

Figure 7.24 shows the corresponding range-Doppler maps.
Again, note the significant spatial-temporal variability of the clutter. Also note the presence of a dominant discrete scatterer near the
end of the trajectory. Finally, Figure 7.25 shows the relative clutter
power as a function of range bin and radar position.

With the new Green's function model, it is possible to readily
ascertain the potential performance gains using channel adaptive
waveforms. If we assume a point target (single range bin target),
we can compute the maximum theoretical adaptive transmitter

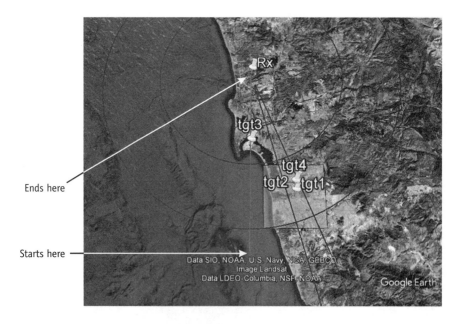

Figure 7.22   Top: IQ simulation using traditional sum over all clutter patches in the
scene. Bottom: IQ simulation using the convoluation of the waveform with the clutter
channel Green's function.

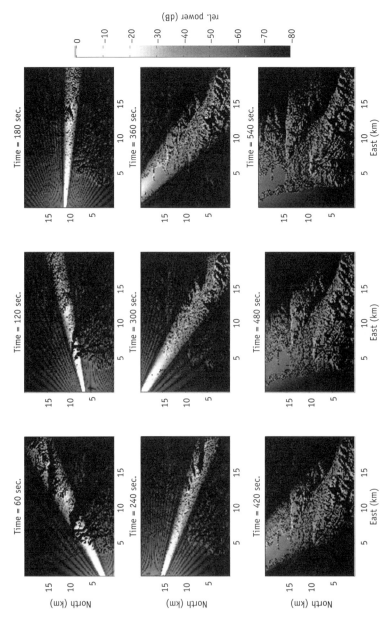

Figure 7.23 Clutter maps from the beginning (upper left) to the end (lower right) of the radar's trajectory. Note how strong the clutter is at the point of closest approach, while towards the end of the trajectory the clutter becomes much weaker and diffuse.

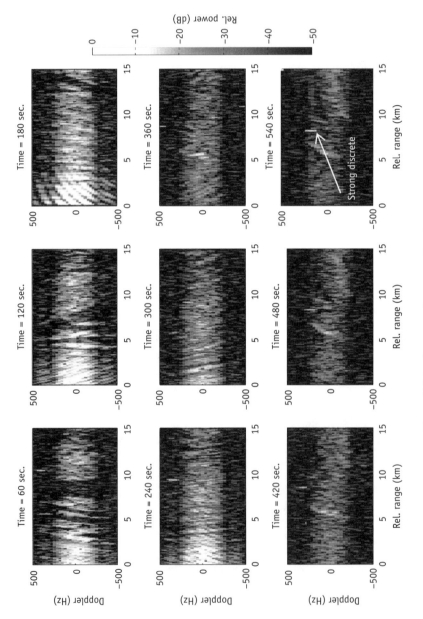

Figure 7.24 Corresponding range-Doppler maps.

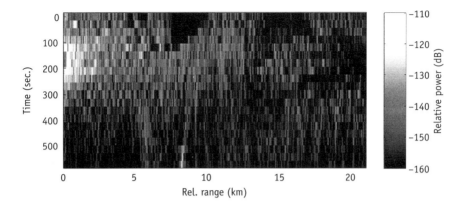

Figure 7.25 Relative clutter power as a function of range-bin and radar position.

waveform gain by calculating the eigenvalue spread of $[(H'_c H_c) + \sigma^2 I]$ where H_c is the discrete (matrix form) Green's function for the clutter channel (fast-time) discussed in detail in Chapter 5.

Figure 7.26 shows the adaptive gain possible as a function of radar (aircraft) position and range. Note that as expected the biggest potential performance gains (~15 dB) are possible in the most extremely strong and heterogeneous clutter environments.

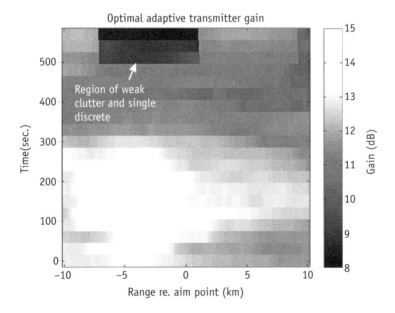

Figure 7.26 Theoretical maximum performance gains using FAR adaptive waveforms as a function of range and radar position.

Interestingly, the region that has relatively weak clutter with a large clutter discrete yields virtually no improvement. That is because the eigenspectrum of $[(H'_c H_c) + \sigma^2 I]$ is basically flat. Thus a maximum bandwidth pulse (or ideally an impulse) is optimum.

This final example is meant to show the importance of having realistic site-specific models of the radar interference environment to support the development of new and advanced optimal waveform and optimal MIMO techniques. It is the highly heterogeneous nature of these environments that makes the optimal waveform approach advantageous. It is very difficult and generally cost prohibitive to use experimental data to support the development of these new optimal waveform techniques since the waveform is part of the solution that cannot be adapted and tested using a static experimental data set collected using a fixed waveform.

References

[1]   https://rfview.islinc.com/RFView/login.jsp.

[2]   Guerci, J. R., and E. J. Baranoski, "Knowledge-Aided Adaptive Radar at DARPA: An Overview," *Signal Processing Magazine, IEEE*, Vol. 23, 2006, pp. 41–50.

[3]   Kang, B., V. Monga, and M. Rangaswamy, "Computationally Efficient Toeplitz Approximation of Structured Covariance under a Rank Constraint," *IEEE Transactions on Aerospace and Electronic Systems*, Vol. 51, 2015, pp. 775–785.

[4]   Wang, P., Z. Wang, H. Li, and B. Himed, "Knowledge-Aided Parametric Adaptive Matched Filter with Automatic Combining for Covariance Estimation," *IEEE Transactions on Signal Processing*, Vol. 62, 2014, pp. 4713–4722.

[5]   Kang,B., V. Monga, and M. Rangaswamy, "Rank-constrained maximum likelihood estimation of structured covariance matrices," *IEEE Transactions on Aerospace and Electronic Systems*, Vol. 50, 2014, pp. 501–515.

[6]   Hao, C., S. Gazor, D. Orlando, G. Foglia, and J. Yang, "Parametric Space–Time Detection and Range Estimation of a Small Target," *IET Radar, Sonar & Navigation*, Vol. 9, 2014pp. 221–231.

[7]   Kang, B., V. Monga, and M. Rangaswamy, "On the Practical Merits of Rank Constrained ML Estimator of Structured Covariance Matrices," in *2013 IEEE Radar Conference (RADAR)*, pp. 1–6.

[8]   Aubry, A., A. De Maio, L. Pallotta, and A. Farina, "Covariance Matrix Estimation via Geometric Barycenters and Its Application to Radar Training Data Selection," *IET Radar, Sonar & Navigation*, Vol. 7, 2013, pp. 600–614.

[9] Monga, V., and M. Rangaswamy, "Rank Constrained ML Estimation of Structured Covariance Matrices with Applications in Radar Target Detection," in *2012 IEEE Radar Conference (RADAR)*, pp. 0475–0480.

[10] Jiang, C., H. Li, and M. Rangaswamy, "Conjugate Gradient Parametric Detection of Multichannel Signals," *IEEE Transactions on Aerospace and Electronic Systems*, Vol. 48, 2012, pp. 1521–1536.

[11] Aubry, A., A. De Maio, L. Pallotta, and A. Farina, "Radar Covariance Matrix Estimation through Geometric Barycenters," in *2012 9th European Radar Conference (EuRAD)*, pp. 57–62.

[12] Abramovich, Y. I., M. Rangaswamy, B. A. Johnson, P. M. Corbell, and N. K. Spencer, "Performance Analysis of Two-Dimensional Parametric STAP for Airborne Radar Using KASSPER Data," *IEEE Transactions on Aerospace and Electronic Systems*, Vol. 47, 2011, pp. 118–139.

[13] Wang, P., H. Li, and B. Himed, "Bayesian Parametric Approach for Multichannel Adaptive Signal Detection," in *2010 IEEE Radar Conference*, pp. 838–841.

[14] Wang, P., H. Li, and B. Himed, "A New Parametric GLRT for Multichannel Adaptive Signal Detection," *IEEE Transactions on Signal Processing*, Vol. 58, 2010, pp. 317–325.

[15] Jiang, C., H. Li, and M. Rangaswamy, "Conjugate Gradient Parametric Adaptive Matched Filter," in *2010 IEEE Radar Conference*, pp. 740–745.

[16] De Maio, A., A. Farina, and G. Foglia, "Knowledge-Aided Bayesian Radar Detectors and Their Application to Live Data," *IEEE Transactions on Aerospace and Electronic Systems*, Vol. 46, 2010.

[17] De Maio, A., S. De Nicola, Y. Huang, D. P. Palomar, S. Zhang, and A. Farina, "Code Design for radar STAP via Optimization Theory," *IEEE Transactions on Signal Processing*, Vol. 58, 2010, pp. 679–694.

[18] Zhang, X., X. Wang, and G. Fan, "Research on Knowledge-Based STAP Technology," in *2009 IET International Radar Conference*, pp. 1–4.

[19] Xue, M., D. Vu, L. Xu, J. Li, and P. Stoica, "On MIMO Radar Transmission Schemes for Ground Moving Target Indication," in *2009 Conference Record of the Forty-Third Asilomar Conference on Signals, Systems and Computers*, pp. 1171–1175.

[20] Wang, P., K. J. Sohn, H. Li, and B. Himed, "Performance Evaluation of Parametric Rao and GLRT Detectors with KASSPER and Bistatic Data," in *IEEE Radar Conference, 2008*, pp. 1–6.

[21] Stoica, P., J. Li, X. Zhu, and J. R. Guerci, "On Using A Priori Knowledge in Space-Time Adaptive Processing," *IEEE Transactions on Signal Processing*, Vol. 56, 2008, pp. 2598–2602.

[22] Gini, F., and M. Rangaswamy, *Knowledge Based Radar Detection, Tracking and Classification*, Vol. 52: John Wiley & Sons, 2008.

[23] Abramovich, Y. I., B. A. Johnson, and N. K. Spencer, "Two-Dimensional Multivariate Parametric Models for Radar Applications—Part II: Maximum-Entropy Extensions for Hermitian-Block Matrices," *IEEE Transactions on Signal Processing,* Vol. 56, 2008, pp. 5527–5539.

[24] Abramovich, Y., M. Rangaswamy, B. Johnson, P. Corbell, and N. Spencer, "Performance of 2-D Mixed Autoregressive Models for Airborne Radar STAP: KASSPER-Aided Analysis," in *IEEE Radar Conference, 2008,* pp. 1–5.

[25] Rangaswamy, M. S., Kay, C. Xu, and F. C. Lin, "Model Order Estimation for Adaptive Radar Clutter Cancellation," in *International Waveform Diversity and Design Conference,* 2007, pp. 339–343.

[26] Morris, H., and M. Monica, "Morphological Component Analysis and STAP Filters," in *Record of the Forty-First Asilomar Conference on Signals, Systems and Computers,* 2007, pp. 2187–2190.

[27] Melvin, W. L., and G. A. Showman, "Knowledge-Aided, Physics-Based Signal Processing for Next-Generation Radar," in *Conference Record of the Forty-First Asilomar Conference on Signals, Systems and Computers,* 2007, pp. 2023–2027.

[28] Gerlach, K. R., and S. D. Blunt, "Radar Processor System and Method," U.S. Patent No. 7,193,558, issued March 20, 2007.

[29] De Maio, A., A. Farina, and G. Foglia, "Adaptive Radar Detection: A Bayesian Approach," in *IEEE Radar Conference,* 2007, pp. 624–629.

[30] Bergin, J. S., D. R. Kirk, G. Chaney, S. McNeil, and P. A. Zulch, "Evaluation of Knowledge-Aided STAP Using Experimental Data," in *IEEE Aerospace Conference,* 2007, pp. 1–13.

[31] Bergin, J. S., D. R. Kirk, G. Chaney, S. McNeil, and P. A. Zulch, "Evaluation of Knowledge-Aided STAP Using Experimental Data," presented at the *2007 IEEE Aerospace Conference,* Big Sky, MT.

[32] Berger, S. D., W. L. Melvin, and G. A. Showman, "Map-Aided Secondary Data Selection," in *IEEE Radar Conference,* 2007, pp. 762–767.

[33] Abramovich, Y. I., M. Rangaswamy, B. A. Johnson, P. Corbell, and N. Spencer, "Time-Varying Autoregressive Adaptive Filtering for Airborne Radar Applications," in *2007 IEEE Radar Conference,* pp. 653–657.

[34] Wicks, M. C., M. Rangaswamy, R. Adve, and T. B. Hale, "Space-Time Adaptive Processing: A Knowledge-Based Perspective for Airborne Radar," *IEEE Signal Processing Magazine,* Vol. 23, 2006, pp. 51–65.

[35] Shackelford, A., K. Gerlach, and S. Blunt, "Performance Enhancement of the FRACTA Algorithm via Dimensionality Reduction: Results from KASSPER I," in *IEEE Conference on Radar,* 2006, p. 8-pp.

[36] Page, D., and G. Owirka, "Knowledge-Aided STAP Processing for Ground Moving Target Indication Radar Using Multilook Data," *EURASIP Journal on Applied Signal Processing,* Vol. 2006, 2006, pp. 1–16.

[37]   Lin, F., M. Rangaswamy, P. Wolfe, J. Chaves, and A. Krishnamurthy, "Three Variants of an Outlier Removal Algorithm for Radar STAP," in *Fourth IEEE Workshop on Sensor Array and Multichannel Processing,* 2006, pp. 621–625.

[38]   Gurram, P. R., and N. A. Goodman, "Spectral-Domain Covariance Estimation with A Priori Knowledge," *IEEE Transactions on Aerospace and Electronic Systems,* Vol. 42, 2006.

[39]   Capraro, G. T., A. Farina, H. Griffiths, and M. C. Wicks, "Knowledge-Based Radar Signal and Data Processing: A Tutorial Review," *IEEE Signal Processing Magazine,* Vol. 23, 2006, pp. 18–29.

[40]   Blunt, S. D., K. Gerlach, and M. Rangaswamy, "STAP Using Knowledge-Aided Covariance Estimation and the FRACTA Algorithm," *IEEE Transactions on Aerospace and Electronic Systems,* Vol. 42, 2006.

[41]   Bergin, J. S., and P. M. Techau, "Multiresolution Signal Processing Techniques for Ground Moving Target Detection Using Airborne Radar," *EURASIP Journal on Applied Signal Processing,* Vol. 2006, 2006, pp. 220–220.

[42]   Ohnishi, K., J. Bergin, C. Teixeira, and P. Techau, "Site-Specific Modeling Tools for Predicting the Impact of Corrupting Mainbeam Targets on STAP," in *IEEE International Radar Conference,* 2005, pp. 393–398.

[43]   Teixeira, C. M., J. S. Bergin, and P. M. Techau, "Adaptive Thresholding of Non-Homogeneity Detection for STAP Applications," in *Proceedings of the IEEE Radar Conference,* 2004, pp. 355–360.

[44]   Rangaswamy, M., F. C. Lin, and K. R. Gerlach, "Robust Adaptive Signal Processing Methods for Heterogeneous Radar Clutter Scenarios," *Signal Processing,* Vol. 84, 2004, pp. 1653–1665.

[45]   Page, D., S. Scarborough, and S. Crooks, "Improving Knowledge-Aided STAP Performance Using Past CPI Data [Radar Signal Processing]," in *Proceedings of the IEEE Radar Conference,* 2004, pp. 295–300.

[46]   Mountcastle, P. D., "New Implementation of the Billingsley Clutter Model for GMTI Data Cube Generation," in *Proceedings of the IEEE Radar Conference,* 2004, pp. 398–401.

[47]   Li, P., H. Schuman, J. Micheis, and B. Himed, "Space-Time Adaptive Processing (STAP) with Limited Sample Support," in *Proceedings of the IEEE Radar Conference,* 2004, pp. 366–371.

[48]   G. R. Legters and J. R. Guerci, "Physics-based airborne GMTI radar signal processing," in *Proceedings of the IEEE Radar Conference,* 2004, pp. 283–288.

[49]   Blunt, S. D., and K. Gerlach, "Efficient Robust AMF Using the enhanced FRACTA Algorithm: Results from KASSPER I & II [Target Detection]," in. *Proceedings of the IEEE Radar Conference,* 2004, pp. 372–377.

[50]   Blunt, S., K. Gerlach, and M. Rangaswamy, "The Enhanced FRACTA Algorithm with Knowledge-Aided Covariance Estimation," in *IEEE Sensor Array and Multichannel Signal Processing Workshop Proceedings,* 2004, pp. 638–642.

[51]   Bergin, J. S., C. M. Teixeira, P. M. Techau, and J. R. Guerci, "STAP with Knowledge-Aided Data Pre-Whitening," in *Proceedings of the IEEE Radar Conference*, 2004, pp. 289–294.

[52]   Rangaswamy, M., and F. Lin, "Normalized Adaptive Matched Filter–A Low Rank Approach," in *Proceedings of the 3rd IEEE International Symposium on Signal Processing and Information Technology*, 2003, pp. 182–185.

[53]   Gerlach, K., "Efficient Reiterative Censoring of Robust STAP Using the FRACTA Algorithm," in *Proceedings of the International Radar Conference*, 2003, pp. 57–61.

[54]   https://lta.cr.usgs.gov/NED 1997.

[55]   TIGER/Line® File Technical Documentation, prepared by the Bureau of the Census, Washington, DC, 1997, http://www.census.gov/geo/www/tiger.

[56]   https://www.mrlc.gov/nlcd2011.php..

[57]   Ulaby, F., and M. Dobson, *Radar Scattering Statistics for Terrain*, Norwood, MA: Artech House, 1989.

[58]   Ruck, Barrick, et al, *Radar Cross Section Handbook,* Plenum Press: New York, 1970.

[59]   Ayasli, S., "SEKE: A Computer Model for Low Altitude Radar Propagation over Irregular Terrain," *IEEE Transactions on Antennas and Propagation*, Vol. 34, No. 8, August 1986.

[60]   Billingsley, J. B., "Exponential Decay in Windblown Radar Ground Clutter Doppler Spectra: Multifrequency Measurements and Model," Technical Report 997, MIT Lincoln Laboratory, Lexington, MA, July 29, 1996.

[61]   Techau, P. M., J. S. Bergin, and J. R. Guerci, "Effects of Internal Clutter Motion on STAP in a Heterogeneous Environment," *Proc. 2001 IEEE Radar Conference*, Atlanta, GA, May 1–3, 2001, pp. 204–209.

[62]   Techau, P. M., J. R. Guerci, T. H. Slocumb, and L. J. Griffiths, "Performance Bounds for Hot and Cold Clutter Mitigation," *IEEE Transactions on Aerospace and Electronic Systems*, Vol. 35, October, 1999, pp. 1253–1265.

[63]   Bergin, J. S., P. M. Techau, W. L. Melvin, and J. R. Guerci, "GMTI STAP in Target-Rich Environments: Site-Specific Analysis," *Proc. 2002 IEEE Radar Conference*, Long Beach, CA, April 22–25, 2002.

[64]   Ohnishi, K., J. S. Bergin, C. M. Teixeira, and P. M. Techau, "Site-Specific Modeling Tools for Predicting the Impact of Corrupting Mainbeam Targets on STAP," *Proceedings of the 2005 IEEE Radar Conference*, Alexandria, VA, May 9–12, 2005.

[65]   Bergin J. S., and P. M. Techau, "High Fidelity Site-Specific Radar Simulation: KASSPER Data Set 2," *ISL Technical Report ISL-SCRD-TR-02-106*, May 2002.

[66]   Bergin, J. S., P. M. Techau, C. Teixeira, and J. R. Guerci, "STAP with Knowledge-Aided Data Pre-Whitening," *Proceedings of the 2004 IEEE Radar Conference*, Philadelphia, PA, April 2004.

[67]   Ward, J., "Space-Time Adaptive Processing for Airborne Radar," *Lincoln Laboratory Technical Report 1015,* December, 1994.

[68]   Leon-Garcia, A., *Probability and Random Processes for Electrical Engineering,* Reading, MA: Addison-Wesley Publishing Company, 1994.

[69]   Zito, R. R., "The Shape of SAR Histograms," *Computer Vision, Graphics, and Image Processing,* Vol. 43, 1988, pp. 281–293.

[70]   Pillai, S. U., H. S. Oh, D. C. Youla, and J. R. Guerci, "Optimum Transmit-Receiver Design in the Presence of Signal-Dependent Interference and Channel Noise," *IEEE Transactions on Information Theory,* Vol. 46, No. 2, March, 2000.

[71]   Guerci J. R., and S. U. Pillai, "Theory and Application of Adaptive Transmission (ATx) Radar," *Proceedings of the Adaptive Sensor Array Processing Workshop,* MIT Lincoln Laboratory, March 10–11, 2000.

[72]   Bergin, J. S., P. M. Techau, J. E. Don Carlos, and J. R. Guerci, "Radar Waveform Optimization for Colored Noise Mitigation," *Proceedings of the 2005 IEEE International Radar Conference,* Alexandria, VA, May 9–12, 2005.

[73]   Bergin, J. S., and P. M. Techau, "An Upper Bound on The Performance Gain of an Adaptive Transmitter," ISL Technical Note ISL-TN-00-011, Vienna, VA, August, 2000.

[74]   Bergin, J. S., P. M. Techau, and J. E. Don Carlos, and J. R. Guerci, "Radar Waveform Optimization for Colored Noise Mitigation," *Proceedings of the Third Annual Tri-Service Waveform Diversity Workshop,* Huntsville, AL, March, 2005.

[75]   Van Trees, H. L., *Detection, Estimation, and Modulation Theory,* John Wiley and Sons, Inc. New York

[76]   Haykin, S., "Cognitive Radar: A Way of the Future," *IEEE Signal Processing Magazine,* Vol. 23, No. 1, Jan. 2006.

[77]   Bergin, J. S., J. R. Guerci, R. M. Guerci, and M. Rangaswamy, "MIMO Clutter Discrete Probing for Cognitive Radar," in *IEEE International Radar Conference,* Arlington, VA, 2015, pp. 1666–1670.

[78]   Guerci, J. R., *Cognitive Radar: The Knowledge-Aided Fully Adaptive Approach,* Norwood, MA: Artech House, 2010.

[79]   Bell, K. L., J. T. Johnson, G. E. Smith, C. J. Baker, and M. Rangaswamy, "Cognitive Radar for Target Tracking Using a Software Defined Radar System," *Proceedings of the 2015 IEEE Radar Conference,* Arlington, VA, May 10–15, 2015.

[80]   Guerci, J. R., "Optimal and Adaptive MIMO Waveform Design," in *Principles of Modern Radar: Advanced Techniques,* W. L. Melvin and J. A. Scheer (eds.), Edison, NJ: SciTech Publishing, 2013.

[81]   Kay, S., "Optimal Signal Design for Detection of Gaussian Point Targets in Stationary Gaussian Clutter/Reverberation," *IEEE Journal of Selected Topics in Signal Processing,* Vol. 1, No. 1, 2007, pp. 31–41.

[82] Bergin, J. S., et al., "A New Approach for Testing Autonomous and Fully Adaptive Radars," *Proceedings of the IEEE Radar Conference*, Seattle, WA, May 2017.

Selected Bibliography

Cobo, B., et al., "A Site-Specific Radar Simulator for Clutter Modelling in VTS Systems," *ELMAR 50th International Symposium*, Zadar, Croatia, 2008.

Don Carlos, J. E., "Clutter, Splatter, and Target Signal Model," ISL Technical Note ISL-TN-89-003 Vienna, VA, November, 1989.

Don Carlos, J. E., K. M. Murphy, and P. M. Techau, "An Improved Clutter, Splatter, and Target Signal Model," ISL Technical Note ISL-TN-91-003, Vienna, VA, May, 1991.

Griffiths, L. J., P.M. Techau, J. S. Bergin, K. L. Bell, "Space-Time Adaptive Processing in Airborne Radar Systems," *The Record of the 2000 IEEE International Radar Conference*, Alexandria, VA, May 7–12, 2000, pp. 711–716.

Guerci, J. R., "Cognitive Radar: The Next Radar Wave?" *Microwave Journal*, Vol. 54, No. 1, January, 2011, pp. 22–36.

Johnson, J. T., C. J. Baker, G. E. Smith, K. L. Bell, and M. Rangaswamy, "The Monostatic-Bistatic Equivalence Theorem and Bistatic Radar Clutter," presented at the *European Radar Conference (EuRAD)*, 2014 11th, 2014.

Melvin, W. L., and J. R. Guerci, "Adaptive Detection in Dense Target Environments," *Proc. 2001 IEEE Radar Conf.*, Atlanta, GA, May 1–3, 2001, pp. 187–192.

Techau, P. M., "Degrees of Freedom Analysis in Hot Clutter Mitigation," Proceedings of the 4th DARPA Advanced Signal Processing Hot Clutter Technical Interchange Meeting, Rome Laboratory, NY, August 7–8, 1996.

Techau, P. M., "Performance evaluation of hot clutter mitigation architectures using the Splatter, Clutter, and Target Signal (SCATS) model," Proceedings of the 3rd ARPA Mountaintop Hot Clutter Technical Interchange Meeting, Rome Laboratory, NY, August 23–24, 1995.

Techau, P. M., "Radar Phenomenology Modeling and System Analysis Using the Splatter, Clutter, and Target Signal (SCATS) Model," ISL Technical Note ISL-TN-97-001, Vienna, VA, October 1997.

Techau, P. M., D. E. Barrick, and A. Schnittman, "The Two-Scale Bistatic Rough Surface Scattering Model," Proceedings of the 2nd ARPA Mountain Top Hot Clutter Technical Interchange Meeting, Rome Laboratory, NY, September 27–28, 1994.

Techau, P. M., J. R. Guerci, T. H. Slocumb, and L. J. Griffiths, "Site-Specific Performance Bounds for Interference Mitigation in Airborne Radar Systems," *Proceedings of the Adaptive Sensor Array Processing (ASAP) Workshop*, MIT Lincoln Laboratory, Lexington, MA, March 10–11, 1999.

Techau, P. M., J. R. Guerci, T. H. Slocumb, and L. J. Griffiths, "Performance Bounds for Interference Mitigation in Radar Systems," *Proceedings of the 1999 IEEE Radar Conference*, Waltham, MA, April 20–22, 1999.

Väisänen, V., et al., "An Approach to Enhanced Fidelity of Airborne Radar Site-Specific Simulation," Proceedings of SPIE Remote Sensing 2008, Remote Sensing for Env ironmental Monitoring, GIS Applications, and Geology VIII, Cardiff, Wales, UK, September 15–18, 2008.

8

Summary and Future Work

MIMO radar is an emerging technology that is beginning to find its way into real radar applications. As we showed in this text, operating in a MIMO mode provides potential theoretical advantages for surveillance and search applications such as airborne GMTI, OTH, and automotive radar. In particular, the MIMO antenna architecture provides improved bearing estimation accuracy in cases when the loss in transmit aperture gain can be overcome or tolerated. This includes applications when system performance is limited by interference as opposed to thermal noise. A good example is the detection of slow-moving targets in clutter in GMTI applications. We also showed in this text that for search applications the loss in antenna gain can sometimes be offset by longer integration times leading to MIMO improvements in thermal noise-limited cases.

One of the main challenges when implementing MIMO systems is the increase in hardware complexity needed to support agile waveforms. Modern AESA radars often provide this complexity for free since they are designed with a capability to transmit spatially diverse waveforms. Adding a MIMO capability to existing traditional single transmit aperture systems, however, can be very challenging, especially if cost is a major system constraint, which is almost always the case with commercial radars.

Another key challenge in implementing MIMO systems is finding waveforms that can be used with practical transmitter hardware. In general, many of the waveform techniques that provide good theoretical performance improvements are not practical since they cannot be transmitted without causing efficiency losses. The most important constraint is that the waveforms often must be constant modulus to work with typical radar hardware employing class C amplifiers. An additional and often more challenging constraint is that the interwaveform phasing must be constrained to ensure that sufficient power is transferred from that input waveforms to propagating modes. This is usually measured with the VSWR metric [1].

The development of practical MIMO waveform techniques that consider both the amplifier constraints and antenna VSWR constraints is not yet mature. Part of the problem is that much of the MIMO work has been performed by engineers and scientists with a signal processing background as opposed to hardware engineers. This divide is likely to be bridged in the future as more MIMO techniques are transitioned into real radar systems, as was the case with the MIMO GMTI radar example presented in Chapter 4. One area of research that would help bridge this gap is the development of antenna modeling tools that are more readily accessible by signal processing engineers and scientist doing the MIMO radar research. An example of some basic analysis tools that help bring the VSWR problem to the forefront were discussed in Chapter 4. The book by Kingsley and Guerci is the first focused attempt at bridg ing the gap between systems and circuits engineers [1].

As discussed in Chapters 5 and 6, there are still significant opportunities to develop optimal MIMO techniques that extend the more commonly employed MIMO antenna with orthogonal waveforms. This will require advanced environmental clutter and interference models that capture the site-specific nature of real-world operating environments that are the most likely to benefit from fine-grained dynamic optimization of the MIMO waveforms. As with the antenna modeling tools discussed above, these types of M&S tools exist; however, they often require specialized expertise in propagation and scattering. Thus they are often difficult to use and apply to signal processing research. A web-based high-fidelity

M&S tool that is commercially available can be found at https://rfview.islinc.com.

MIMO radar is an exciting technology with clear benefits in existing systems and is likely to find new applications in future systems. Unlike radar signal processing technologies such as STAP, which typically only impact the receiver, MIMO techniques require signal processing researchers to take a more hardware-oriented view of radar systems. While this will require more detailed models and a more multidisciplinary approach to research, as was shown in the text, it is likely to open up new ways of operating radars that when done judiciously can open up exciting and new capabilities and systems.

Reference

[1]   Kingsley, N., and J. R. Guerci, *Radar RF Circuit Design*. Norwood, MA: Artech House, 2016.

About the Authors

Joseph Guerci has over 30 years of advanced technology development experience in industrial, academic, and government settings—the latter included a seven-year term with Defense Advanced Research Projects Agency (DARPA) where he led major new technology development efforts in his successive roles as program manager, deputy office director, and director of the special projects office. He is currently president and CEO of Information Systems Laboratories, Inc.

Dr. Guerci, who has a Ph.D. in electrical engineering from NYU Polytechnic University, is the author of over 100 technical papers and publications, including the books *Space-Time Adaptive Processing for Radar*, 2nd ed., (Artech House) and *Cognitive Radar: The Knowledge-Aided Fully Adaptive Approach* (Artech House). He is a fellow of the IEEE for *Contributions to Advanced Radar Theory and Its Embodiment in Real-World Systems*, and the recipient of the 2007 IEEE Warren D. White Award for *Excellence in Radar Adaptive Processing and Waveform Diversity*.

Jameson Bergin is a graduate of the University of New Hampshire with over 20 years of experience developing advanced radar concepts and adaptive signal processing techniques. He has authored or coauthored numerous conference and journal articles on MIMO radar, cognitive radar, space-time adaptive processing, and knowledge-aided processing. He is the instructor of a popular tutorial on MIMO radar that has been presented numerous times at the IEEE Radar Conference. He currently works for Information Systems Laboratories, Inc. and resides in Glastonbury, CT, with his wife Ruth and their four children, Livia, Madelyn, Lucy, and Zachary.

Index

Principles of Radar and Sonar Signal Processing,
  François Le Chevalier

Radar Cross Section, Second Edition, Eugene F. Knott, et al.

Radar Equations for Modern Radar, David K. Barton

Radar Evaluation Handbook, David K. Barton, et al.

Radar Meteorology, Henri Sauvageot

Radar Reflectivity of Land and Sea, Third Edition, Maurice W. Long

Radar Resolution and Complex-Image Analysis, August W. Rihaczek
  and Stephen J. Hershkowitz

Radar RF Circuit Design, Nickolas Kingsley and J. R. Guerci

Radar Signal Processing and Adaptive Systems, Ramon Nitzberg

Radar System Analysis, Design, and Simulation, Eyung W. Kang

Radar System Analysis and Modeling, David K. Barton

Radar System Performance Modeling, Second Edition,
  G. Richard Curry

Radar Technology Encyclopedia, David K. Barton and
  Sergey A. Leonov, editors

Radio Wave Propagation Fundamentals, Artem Saakian

Range-Doppler Radar Imaging and Motion Compensation,
  Jae Sok Son, et al.

Robotic Navigation and Mapping with Radar, Martin Adams,
  John Mullane, Ebi Jose, and Ba-Ngu Vo

Signal Detection and Estimation, Second Edition, Mourad Barkat

Signal Processing in Noise Waveform Radar, Krzysztof Kulpa

Space-Time Adaptive Processing for Radar, Second Edition,
  Joseph R. Guerci

For further information on these and other Artech House titles, including previously considered out-of-print books now available through our In-Print-Forever® (IPF®) program, contact:

Artech House	Artech House
685 Canton Street	16 Sussex Street
Norwood, MA 02062	London SW1V HRW UK
Phone: 781-769-9750	Phone: +44 (0)20 7596-8750
Fax: 781-769-6334	Fax: +44 (0)20 7630-0166
e-mail: artech@artechhouse.com	e-mail: artech-uk@artechhouse.com

Find us on the World Wide Web at: www.artechhouse.com